TWO TRAILS

A HIKER'S STORY OF SURVIVAL AND HIS MOTHER'S STORY OF HOPE

KAREN ZIEBARTH

WITH CONTRIBUTIONS BY **ERIC DESPLINTER**

Mountain House Publishing

©2022 Karen Ziebarth.

All Rights Reserved. No part of this publication may be reproduced, stored in a retrieval system, or transmitted in any form or by any means—electronic, mechanical, photocopy, recording, or any other—except for brief quotations in printed reviews, without the prior permission from the publisher.

Products and services mentioned throughout this book are owned by their respective copyright, patent, and trademark holders. Photographs used in this book are from the personal collections of the author or properly licensed from various stock photo houses.

Disclaimer: This is the author's story. It is not intended to be a guide or how-to manual for a similar trip. No claims are made as to legality, suitability, safety, or efficiency of any route, road, service, establishment, or method of travel. State laws may have changed since this trip was taken, and the author recommends you familiarize yourself with state and local laws where you plan to travel.

Mountain House Publishing

www.MountainHousePublishing.com

Color Hardcover ISBN: 979-8-9869502-1-1

Color Paperback ISBN: 979-8-9869502-2-8

B&W Paperback ISBN: 979-8-9869502-3-5

Kindle ISBN: 979-8-9869502-4-2

Library of Congress Control Number and Cataloging-in-Publication data on file with the publisher.

Publishing and Production by Concierge Marketing Inc.

Printed in the United States of America

10 9 8 7 6 5 4 3 2 1

TWO TRAILS

Contents

Preface vii

Dedication x

Prologue 1

Day One – Saturday, April 6, 2019 3

Day Two – Sunday, April 7, 2019 13

Day Three – Monday, April 8, 2019 21

Day Four – Tuesday, April 9, 2019 35

Day Five – Wednesday, April 10, 2019 51

Thank You to All of the Search and Rescue Teams 75

A Few "Whys" of Being Part of a Search and Rescue Team 77

Rescue Crews 83

Epilogue 95

Rescue Day 103

About the Author 105

Preface

The beauty of the outdoors offers peace and solitude for some people and opportunities to challenge physical endurance for others. For many, it does both. Hiking adventures for my family began in 2016, when I told my oldest son, Eric, I wanted to see Yosemite National Park and asked if he'd like to go with me. He agreed and called the following week: "So, I've got the hikes picked out that I want to go on. How about you?" Hmmmmmmm …hikes? I hadn't thought about that. I thought we'd be doing the scenic-drive method of vacationing, but, hey, I'm game. Hiking could be fun.

Indeed, it was. The trip sparked an even greater appreciation for nature in me and both my sons, Eric and Tim. Eric and I have set out on a journey to visit every national park in the United States. Yosemite was our first. It seemed majestic to us, and we thought subsequent trips would pale in comparison. We've been proven wrong, as each park has its own unique and breathtaking beauty. Yosemite, Zion, Yellowstone, Grand Teton, Kenai Fjords, Katmai, Lake Clark, Wrangle-St. Elias, Glacier Bay, Joshua Tree, Glacier, Petrified Forest, Mesa Verde, Great Sand Dunes, Black Canyon of the Gunnison, Arches, Canyonlands, Badlands, Wind Cave, Theodore Roosevelt, Acadia, Isle Royale, and Voyageurs have been ticked off our list. Eric now lives in California, so he's been known to cheat a bit and get ahead of me on solo trips; however, a park is not added to the official list until we've been there at the same time.

Since moving to California, Eric has also taken advantage of hiking in the San Gabriel Mountains on many occasions with friends. On April 6, 2019, he went hiking with his friends Gabrielle, Charles, Nicole, Ken, Nalinee, and Jon. On this trip, however, he didn't return as scheduled. This is Eric's story of the physical and mental challenges of surviving for five days while working to navigate out of the mountains. It is also the story of his family praying him home.

—Karen Ziebarth, Eric's mom

Thanks to Willie Hunt (right) and Sonny Lawrence (left) for teaching me the skills to safely navigate the canyon and accompanying me on my return trip to the Cucamonga Canyon.
– Eric

This book is dedicated to my children, Nikki, Eric, and Tim, who will always be my everything. The world is a better place with you in it.

For Tara, who always saw the best in every day.

For my grandchildren, Boomer, Dysen, and Renaud.
Always remember… Namma says you're perfect.

In remembrance of:

My parents, Kenneth and Frances Ziebarth.
Thanks, Dad, for always teaching by example and living your life with integrity.

My grandfather, Francis Gurneau, the most gentle soul I've known and my guide as I was growing up.

Thank you to the following individuals for their contributions in writing this book:

Jeff Lehman, San Bernardino Search & Rescue, for the countless questions answered and meetings arranged.

Frank Kams, San Bernadino Sheriff's Emergency Services Coordinator

Art Fortini, Operations Leader, Sierra Madre Search & Rescue

Donna Newlin, San Bernardino County Sheriff's Search & Rescue, Inland Empire SAR Council

Photo by Gregory Boratyn
Greg Boratyn Photography, scenesfromlight.com

Prologue

On April 7, 2019, I was working at home when I received a text from a number I didn't recognize.

"Is this Karen, Eric DeSplinter's mother? Last night he and a coworker did not return from a hike. Search and rescue efforts started around 10 p.m. and are continuing right now. The hike was from Ice House Canyon Trailhead to Cucamonga Peak. We last saw them at Ice House Canyon Saddle, 3.5 miles from the trailhead. The phone number of the officer in charge of the search and rescue efforts is 909-501-****."

My first reaction was to wonder how the sender had found my phone number. Then I re-read the message much more slowly, and the words "search and rescue" hit me hard.

Day One – Saturday, April 6, 2019

Eric: To Cucamonga Peak

My favorite part of moving to Ontario, California, is living so close to the mountains. I had spent six years in St. Louis, Missouri, working long days and devoting most of my evenings to home improvement projects. In California I began spending more of my free time outdoors with others in the office who enjoy hiking.

Today I was planning to hike to the top of Cucamonga Peak with two of my coworkers, Charles and Gabrielle. We were joined by Charles' wife, Nicole, and three additional friends, Ken, Nalinee, and Jon. It was a chilly morning, but I knew we'd warm up in no time. We had 4,300 feet of vertical gain to go, and it would be more than 11 miles up and back. It was going to be a good day.

We had planned to hike to the Ice House Saddle and take a short break before heading the rest of the way to Cucamonga Peak. The trail up to Ice House Saddle was free of snow, and we could see that the remainder of the trail would require our crampons. Ken, Nalinee, and Jon had not brought theirs, so they did not continue beyond Ice House Saddle. After Charles, Nicole, Gabrielle, and I donned our crampons, we continued up the trail toward Cucamonga Peak. A few hundred yards after the saddle, Charles and Nicole decided to turn around because the terrain on that day was a little beyond their comfort level. We planned to meet back at the trailhead and find a place for dinner at the end of the day.

Gabrielle and I continued hiking toward Cucamonga Peak. At first we had obvious footprints to follow. The footprints were slightly gaining elevation as we skirted around Big Horn Mountain. The further we went, the less pronounced the established path became. Eventually the footprints stopped. We checked Gabrielle's GPS app, Gaia, and discovered that we had climbed a few hundred feet higher than the actual trail. The path was getting a little difficult, so we decided to descend and try to find the trail again. We walked downhill, with Gabrielle now above me in case she slipped. My footing was secure, and I would be able to stop her if she started to slide. Eventually, we descended to the trail elevation and could once again see the actual trail. The snow and ice on short portions had melted, which helped us follow the trail as we skirted around Big Horn Mountain.

The rest of the hike to Cucamonga Saddle was relatively easy because others had walked on the path and created a flat, worn route for us to follow without exerting much energy. We crossed paths with individual hikers on two occasions. They both said we shouldn't have a problem making it to the summit and that the view was spectacular. One hiker said he had also followed the same tracks we had previously followed—the ones that had taken us higher than the trail. He also warned us that some of the tracks on the way to Cucamonga Peak might be misleading, since he had lost the trail multiple times.

Ice House Saddle

Switchback snow-free on the return trip to Cucamonga Peak

When we reached Cucamonga Saddle, I decided to eat the rest of my food: carrots and an energy bar. I assumed the peak was going to be windy, and I didn't want to spend time eating up there. Plus, I needed extra energy as we were about to hike the final 1,000 feet of elevation gain via steep switchbacks to the peak.

We started walking on the first switchback. The ice was harder than it had been between Ice House Saddle and Cucamonga Saddle. This area was exposed to direct sunlight, which would melt the snow. Then the melted snow would freeze again, resulting in a hard ice surface. I turned to start the second switchback. As I turned, I saw Gabrielle walking up the first switchback, when she suddenly slipped. She started sliding down the mountain. I thought she was going to stop shortly after falling, but she kept sliding and quickly started picking up speed. I kept yelling to her to self-arrest. About 100 feet down the mountain, she tried. Her ice axe stuck in the ice, but she couldn't hold on. She fell another 200 feet before she was stopped by a large manzanita bush protruding above the snow. It was like watching her fall in slow motion. I could see she was heading for shrubbery that would stop her. Gabrielle didn't know that, and it must have been horrifying for her.

I yelled down to her, asking if she was okay. She said she was, but I knew she had been shaken. I hollered for her to stay put and I would come to her. I said we were done for the day and we would head back to the car. I started descending, keeping an eye out for Gabrielle's phone that she had lost in the fall. Although we had been using Gabrielle's app on her phone for our trail map, I wasn't concerned that I couldn't find it because we were planning to climb back to the trail and head back down the mountain the same way we had come in. On my descent, I was able to grab Gabrielle's ice axe.

Snow- and ice-free descent close to where Gabrielle slipped.

Looking from bottom up at the location where Gabrielle fell.

When I reached Gabrielle, I noticed that she was guarding her hands and appeared to be in pain. She had broken several fingernails, and skin was ripped off the palm of her hand. I saw blood on the ice near where she was sitting. As she moved, I saw that her left pant leg was torn a couple of inches, and her leg was cut. She also had blood on her shirt near her oblique. Her injuries were not life-threatening, so we decided to set back up the mountain to return to the point where she had slipped.

Our plan was to hike back up to the trail and egress the way we had come in. While maneuvering back up to the trail, I tried to reassure Gabrielle by reminding her that I was right next to her and we would get to the trail again together, one step at a time. She was moving extremely slowly. It seemed like she had lost confidence in her ability to walk on the ice, which would be expected after taking such a significant fall. I was also concerned about her potentially falling again as we skirted back around Big Horn. If she fell there, it would be the opposite way of home, and I didn't want that to happen.

I looked around for a safer path. I noticed that another 200 feet down the mountain, the snow and ice was gone and there was a wide-open valley. The valley led to the south, which would end in the foothills. I asked Gabrielle if she would rather walk down to the valley so she wouldn't have to be on the snow and ice anymore. I thought this was a good option because she had the strength and balance to hop from boulder to boulder through the valley. She was onboard with that idea. We walked down to the valley and removed our crampons. I tried texting Charles to tell him our plan and ask him to pick us up in the foothills. Unfortunately, I didn't have cell reception, and my text never went through.

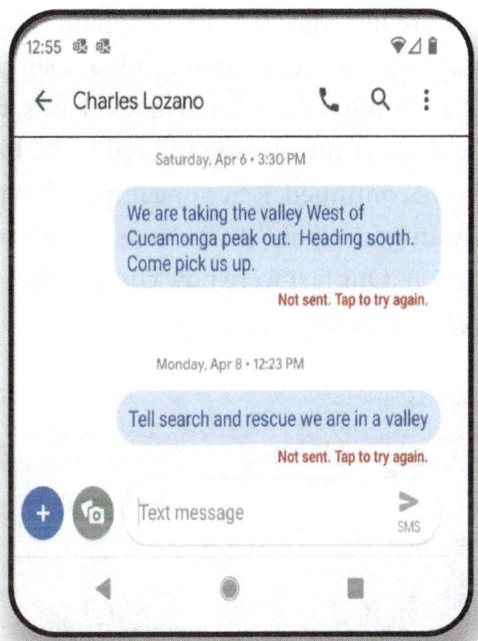

Screenshot of text messages that Eric tried to send to Charles.

It was now mid afternoon, and we started descending through the valley as fast as we could. We figured we had plenty of time to get to the foothills. The valley was wide, with large boulders, limited vegetation, and no water. The terrain was easy to traverse, and we descended about a mile until it became densely forested. At one point, I saw water coming from the mountain and I stopped to take in the view. It was cool to see the point where the water starts coming out of the mountain. Although it was a small amount of water, I became concerned at that point, knowing that if the water became more significant, it would slow our progress. I also thought about the canyon potentially narrowing, but it was still fairly wide, and I had been in plenty of similar canyons that I could walk through. We continued downstream. The vegetation was getting dense, which made our progress slow, as we had to travel away from the stream. We wanted our feet to stay dry, so we didn't try to walk through the stream.

We kept crossing the stream, hopping along rocks and attempting to stay dry. As I was crossing a log, I slipped and landed horizontally with the log on my lower back. I was concerned how that impact would affect my already injured back, but I kept going without pain. As we continued, it became impossible to keep our feet dry. We simply wanted to make progress as quickly as possible and get out of the canyon. Our feet were now completely wet, so we saved time by walking through the water.

The farther we walked, the narrower the canyon became. The water flow also got heavier. The sun was going down, which was not going to work in our favor. Although we didn't want or need another obstacle, we encountered one—a waterfall. We stopped to figure out our next move and found that the waterfall could be avoided by descending from a ledge about 12 feet high on the right side of it. Uneven rocks and boulders at the bottom would make it risky to jump from the ledge. I thought we were close to the end of the canyon, and I wanted to make the jump. Gabrielle wasn't comfortable doing that, so we decided to go back upstream a little bit to find a place to sleep.

We found a sandy area and quickly got ready for what would be our landing spot for the night. I cracked a chem light so we had some light to work with at our campsite. We decided not to start a fire because the vegetation was so thick that we were worried about starting a forest fire. We pulled out my emergency blanket and put on all of our layers of clothing. We put our ice axes next to our sleeping position for defense in case we were visited by a predator during the night. Our survival kit had an emergency blanket, but it was not thick enough to be comfortable. It was too cold for our limited layers of clothing.

That night two helicopters passed directly over us. They were flying perpendicular to the canyon and had almost no time to see us. We figured the helicopters might have been search-and-rescue looking for us and thought perhaps Charles and Nicole had notified them. As I stretched out and tried to get some rest, I thought about how much Sunday was going to suck because I wasn't going to get a good night's rest, and tomorrow was going to be a long day. I had work that I needed to get done before Monday morning, and I wasn't looking forward to trying to complete that work after being deprived of a full night of sleep. In the morning we would head to the foothills, contact Charles to pick us up, and then I needed to get home to shower so I could go to the office. Had I known what the next four days were going to bring, I would have welcomed that scenario.

Day Two – Sunday, April 7, 2019

Eric: In the Canyon

When we woke up on Day Two, I had sharp low back pain, which I had been dealing with earlier in the week. This was not how I had hoped to start the second day of our journey. Gabrielle asked if I was sure we were traveling the right way. I said yes and pulled out a compass to reassure her. To my surprise, the compass said downstream was due north. I pulled out a second compass, which confirmed what the first one showed. I was totally confused. There was no way that could be accurate. I swore I knew which direction we were traveling but didn't have a map to confirm it. As we descended on Day One, the sun was setting over the right side of the canyon.

Everything was saying we were traveling south, except for the compasses. We decided to wait until the sun came over the canyon wall to confirm our direction. As the sun rose over the right side of the canyon, we knew we were facing north. I apologized for messing up, and we set off hiking back upstream. Our goal was to get close to where Gabrielle fell and then return to the trailhead the way we had come the previous day.

Because my back hurt, I used my trekking poles to assist me in getting back up the canyon. I tried to avoid bending below obstacles as much as possible. Gabrielle helped to push me up over larger obstacles. When we got to the point where the water started, the location looked different from what we recalled seeing the day before. This was a little concerning, as we thought maybe we had missed an offshoot into a different canyon. This could have happened because portions of the canyon were densely forested.

We continued walking up the dry portion of the canyon looking for things we recognized. Then we came to another stretch of water. We didn't recall the water starting, stopping, and starting again during our descent. This also confused us. I pulled out the compass to check our direction of travel. The compass indicated that we had been traveling north today. How could that be? Were we in a different canyon? We didn't remember the area we were in at all, but we also weren't trying to remember what we were walking by on Day One because we had just thought we would walk out the south end of the canyon that day. We looked around for clues, and I found a footprint. I told Gabrielle to come over to me. We looked at her shoe and confirmed it was her footprint. We then knew we were in the right canyon. I suspected we had been going in the right direction on Day One and that the canyon had just happened to do a quick 180-degree turn where we stopped. Of all the places for us to have camped the first night, what were the chances of it being in an area where the watercourse turned north? Although that possibility seemed unlikely, I couldn't think of any other explanation.

We were both tired from ascending all day, traveling about two miles with an elevation gain of approximately 3000 feet, only to be near the spot where we began our journey the day before. We decided to find an open spot in the valley to set up camp. We collected firewood, started a fire, and cleared a sleeping area. When we finished preparing for our second night in the canyon, we saw a helicopter in the distance and signaled to it with the signal mirror. The helicopter flew over to our position. I waved my trekking poles over my head, and Gabrielle waved the emergency blanket. The helicopter circled our position and flew away. We didn't know why it flew away, but we assumed that it was going to get the rescue team. We were elated. I commented that I had wasted my time making a bed. We talked about what we were going to eat at McDonald's when we got back. We waited and waited; and the longer we waited, the quieter the world seemed. The helicopter never returned. We were confused and didn't understand how the search-and-rescue team couldn't have seen us. We sat around the fire for a while until we needed to sleep. The night was very cold, thanks to a breeze from the north.

Karen: The Home Front, 11:36 a.m.

My Sunday morning began as most do for me—working from home. I provide closed captioning services for hearing impaired students and live television programs using a speech-generated program. Covering sports is my favorite, so I was starting my day on a high note. At 11:36 a.m., I received a text from a number I didn't recognize.

"Is this Karen, Eric DeSplinter's mother? Last night he and a coworker did not return from a hike. Search and rescue efforts started around 10 p.m. and are continuing right now. The hike was from Ice House Canyon Trailhead to Cucamonga Peak. We last saw them at Ice House Canyon Saddle, 3.5 miles from the trailhead. The officer in charge of the search and rescue efforts number is 909-501-****."

As I read the words, nothing was registering in my head. All I could think about was how someone got my number and knew Eric. I read the message again, much more slowly this time. I remember reading "we last saw them," but that still didn't jolt me. It was when I re-read "search and rescue" that I began to panic. As I said out loud, "No, no, no, no, no, no, no, no," I realized my microphone had picked up my voice, and those words danced across my television screen.

I immediately sent a message that I needed to get off air so I could leave right away. Waiting for my backup for that last 15 minutes was grueling.

I knew I had to travel from Iowa to California right away, but I couldn't decide what to do first. I knew I needed to pack, but I also wanted to get to my daughter, Nikki, and my other son, Tim, so they could be told face to face that Eric was missing. I preferred they not find out through a phone call. The next few minutes consisted of wandering back and forth between my bedroom to pack and the office to pack work equipment and getting sidetracked with trying to get an airline ticket. I was frustrated trying to prioritize what needed to be accomplished first. Securing a ticket should be the priority, but I didn't know how many to get. Would Nikki want to fly out immediately? Would Tim want to go? I wouldn't know until I could tell them what had happened.

I shifted my focus to getting to Nikki and Tim, so I went to the garage and got in the car, but I felt nauseated and my leg shook as I pushed the brake pedal. I finally decided it would be best to call Nikki and tell her on the phone that her brother was missing from a hiking trip. She would know the severity immediately. There was no way Eric was missing unless something was terribly wrong. She immediately figured that out, and I could feel her fear through the phone.

Within 30 minutes Nikki was at my house. I had made no progress by the time she arrived, but her presence alone helped. The moment she walked through the door, she had a plan and hit the ground running. She had my grandchildren—Boomer, Dysen, and Renaud—gather my suitcase, clothes, and essentials. I, on the other hand, was grabbing hiking gear and my work equipment.

"Why do you want your work equipment?" Nikki asked, appearing puzzled.

I didn't know how long I was going to be in California. Were we going to find him right away? Was he going to be okay? One of my fears was that he was injured and paralyzed. If that were the case, Eric would be on a long journey to recovery, and I would stay until he was back up and on his feet. That scenario was better than someone telling me he wasn't alive. That had to be an impossibility, and I tried not to let that thought creep in. In the meantime, Nikki continued to pack and picked up where I left off with securing airline tickets.

"Tim is going with you," she said.

Yes, good plan, but I hadn't told him yet, so here we were again with me not wanting anyone to find out on the phone that their brother was missing. I knew Tim was working and couldn't talk to me, so I had to finally resort to sending a text message. I wanted to offer the choice for him to fly with me, but the decision had to be his. I did, however, secretly hope he would want to go. Without question, he was on board.

I was nervous that Eric might be seriously hurt, and I was at a loss for the right words to say to my two other children. When those three are all together, they make the most awesome team. They're quite entertaining, and I love to listen to them play off one another's humor. This ordeal was going to be hard if it didn't end well. I didn't want to think like that, but I knew something had to be horribly wrong because Eric hadn't made it back. I don't remember much of the trip to the airport other than the fact that Nikki drove 100 miles per hour so we could make the 3:30 p.m. flight.

Tim sat next to me on the flight from Des Moines to Chicago. We were doing well on that leg. Tim and I immediately put in our earbuds, like we do every time we fly together, trying to find something to watch to make the time go by more quickly. There wasn't a lot of time on the layover, just enough to get to our next gate and to confirm the rental car at LAX. We boarded the plane. Tim ended up sitting toward the front, and I was a couple of rows from the back. We were on the tarmac for a couple of hours, waiting for them to fix a mechanical problem. The sun began to set. As it got dark, I started to cry and couldn't stop, no matter how hard I tried. I just wanted Eric to be okay. It was going to be dark out there for him soon. Where was he? Had search and rescue made any progress in finding him? I wanted to call the search-and-rescue command center and get an update. Instead, we sat on the tarmac in a dark plane almost 2,000 miles away. I think the young lady sitting next to me thought I was a bit unstable. Hiding one's tears is difficult when you're in close quarters. I tried to compose myself to be able to have an adequate conversation, but that didn't go well either. I had thoroughly used every tissue I had, and the woman was clearly uncomfortable. Rightly so. She leaned away from me, and I sat and waited in darkness for the plane to finally take off.

We landed at LAX, got a rental car, and headed to Eric's apartment. Eric's co-workers had texted me that they were worried about Eric's cat, DJ, so Tim and I stopped to see if we could get into the apartment to check on him. If Eric had been rescued while we were en route to California, his first question was going to be "Did you stop to take care of DJ?" so we thought we'd be a step ahead in the game. We called the apartment complex, who told us to call the Humane Society, who told us to call the police department. Although Tim's last name on his driver's license matched Eric's, nobody would allow us entry to the apartment nor assist us with checking on DJ, so we headed for the command center where they were coordinating rescue efforts.

We arrived at 2 a.m. to a dark fire station in Mt. Baldy, where they were supposed to be organizing the efforts to look for Eric and Gabrielle. There was nobody there. The last message I had received from Charles and Nicole was that search-and-rescue helicopters were searching throughout the night, but the command center was like a graveyard. I couldn't understand why. It was bitter cold and the wind was whipping. I got out of the car and stood in the cold darkness. Eric had to be freezing, and I wanted to at least be in the same conditions he was, but I couldn't do it. I simply couldn't take it anymore, so I got back into the car and turned the heater on. Although it makes no sense, I felt I was deserting him by not enduring the cold as he was.

Sitting in the car with Tim, I tried to picture what else Eric was having to go through now. Had his body temperature dropped lower because he was injured? Was he in shock? Was Gabrielle hurt? Something simply had to be wrong, because Eric doesn't just not return from a hike on time unless there is a serious issue. He thrives on being challenged and problem solving. He would see this as a puzzle and had to be putting the pieces together to come up with a solution.

The one thing nobody could change was the weather, and it was wicked at night. I had packed layers like we had on our trip to Alaska. I got out of the car again. Although I was four layers deep with my regular hiking gear, long-sleeved base layer, long-sleeved merino shirt, fleece sweatshirt, and waterproof GORE-TEX jacket, I was still freezing, visibly shaking. The wind was unbearable. If the wind would just stop blowing and the sun would come up, Eric would at least be warm. Envisioning him lying on rocks somewhere unable to move, shaking from the frigid temperatures and wind, was distressing. He was out there in the pitch-black dark and freezing cold, with no choice but to gut it out, and I couldn't even sit outside a vehicle until the sun came up.

Day Three – Monday, April 8, 2019

Eric: In the Canyon

We thought the easiest route to safety would be to go back down the canyon. Ascending was physically tiring and we would have to put on our alpine gear once we reached the snow, which would put us at risk of another fall. We set out down the canyon to go back south. About two minutes into our hike, Gabrielle said she was tired. Her body had taken a beating. She asked if it might be better to wait in the wide canyon area to be found. My feet were blistered and sore, and my back hurt. After discussing our options, we agreed that survival sometimes needs to be lazy, so we decided to stay in place for the day and hope the rescue team came back, since we'd had a near miss with the helicopter the day before. Resting would also give our bodies some time to recover so we would be in better condition to move the next day if we weren't found. We waited the entire day but only heard and saw commercial jets.

We were able to get some rest during the warmth of the day, so it wasn't a complete waste. But it was disheartening because we thought we had a good chance of being found that day, and we didn't understand how the helicopter the previous day had not seen us or why they didn't come back.

Knowing we were in for another cold night, we set up a new camp about 200 feet south of where we had camped on Day Two, hoping that the new location would provide a little more protection from the wind. We started a fire but didn't have a decent place to lie down, as the area was rocky. Sleeping on rocks beats up a body quickly, so we slept sitting up near the fire.

Karen and Tim: At the Rescue Command Center

It was still dark when the first signs of life arrived at the command post. I gave the gentleman about 60 seconds to turn on the lights and then I went to check in with him, a man who appeared to be in his 50s with gray hair and a long mustache. Learning about the rescue plan at least would give us some assurance that progress was being made. I told the man that as soon as the sun came up, Tim and I were going to walk the trail to the place where Eric was last seen. He looked at me and sternly let me know that my plan was not a good idea. That was not the response I wanted, and I told him we were going to walk the trail. But for now, I figured this could be a long journey, so we needed to table that idea for a short while.

Slowly—it seemed excruciatingly slowly rescue workers arrived. I didn't ask how many people would be participating in the rescue, but it appeared to be 40 to 50. As the sun came up and the temperature began to warm ever so slightly, Tim and I started to feel a bit of hope, but it seemed as though everyone was moving in slow motion. We didn't understand what they were waiting for; I wanted them to move faster and head out to find Eric. It was mid-morning by the time the first team set out. They would break apart in small groups of three to six while remaining with members of their own rescue team with whom they had trained. They all presented themselves professionally and with confidence. Tim and I were grateful to know that several competent crews were actively searching and that

Eric and Gabrielle weren't alone out there. They may not have been located yet, but people were looking for them.

The sun warmed us as the day went on. I felt much better knowing that at least Eric and Gabrielle weren't shaking from the cold during the day, but time moved slowly. Cell reception was unreliable on the mountain, which made it difficult to get updates to and from Nikki. She had to run the business in my absence since I was unable to communicate consistently on the mountain. I was grateful that all of my work worries were in her capable hands. Nikki can multitask better than anyone I know, and I knew she could problem-solve without having to reach me. I also felt guilty that so much had been dumped on her. I thought about her often throughout the day. Did every minute seem like an hour for her as it did for us, or did time pass more quickly with so much added to her plate?

Tim and I needed something to do, so although search and rescue didn't want us to walk the trail to Ice House Saddle, we did it anyway. Tim and I had hiked many trails together, and we wanted to see for ourselves the terrain that had been described as treacherous. It was important for us to know exactly what Eric had seen because we thought it might help us get an idea of why he wasn't getting back to us. Although hiking the trail wasn't doing anything to help, it gave us something to occupy our minds, and we felt like we were doing *something* positive instead of sitting around. It was just nice to be moving and walking the same path we knew Eric and Gabrielle had taken.

The first part of the trail was breathtakingly beautiful. Wind rustled the leaves in the woods, offering an indescribable sense of peace. We could hear the water rapidly babbling over the rocks. I knew how much Eric would have loved this part of the trail. He loves all parts of the hikes but gravitates toward the water. With each step I took, I wondered if I was stepping in the same spot where Eric had just been. Many others likely had traveled that route since Eric had been on it, but I wanted to make some connection that would help me feel like he was still alive, even if it was as insignificant as stepping in the same spot where he had been.

Just past the trailhead to Cucamonga Peak.

The trail was well maintained and wasn't an extremely tough route. Parts of the hike were strenuous with a steeper incline, and that's when our heart rates increased, but I just stopped to rest when necessary. Tim, on the other hand, did a mad charge to the top, leaving me in the dust. When the trail split, I ended up going to the right, which I found was not the choice I should have made. I spotted the trail to my left again and had to scramble through an area half the size of a football field, filled with large and small boulders. When I finally reached Tim at the saddle about 20 minutes later, we wanted to go farther up the trail to Cucamonga Peak, but there was too much snow to hike farther without crampons. We stayed at the saddle a while so we could be where Eric had last been seen by his friends.

Tim and I could hear rescue workers hiking up the back side of the mountain. I didn't want to leave until I could see them and talk to them. I envisioned them coming up over the crest and telling us that Eric and Gabrielle had been found. As I sat on a rock and listened to the rescue workers' voices, I imagined them coming into sight hollering, "We found them!" That didn't happen but pretending that everything was fine comforted me in that moment.

Half a dozen rescue workers came over the crest, one by one, and walked over to talk to us. These people amazed me. I was told they were volunteers donating their time for this search, as they do for many others. That's admirable. We talked a bit longer and then parted ways as we started back down to the trailhead, hollering for Eric and Gabrielle as we left. If Eric was lying out there injured, I hoped at least he could hear our voices so he would know we were coming to find him. I hollered, "Eric," and it warmed my heart when Tim would holler, "E." Only Tim calls Eric that. I wished I could see them face to face again and hear Tim call him E.

Tim and I returned to the village and took up residence on a rock wall that was the home to a flagpole in front of the fire station. That wall became our designated spot to sit and wait. Sometimes we talked and tried to figure out what we could do next to help, and sometimes we sat silently with our thoughts. I was glad Tim was with me. My mind was occupied by playing out every scenario of how this would end. If the outcome was not good, I wanted to be prepared. Sometimes people stopped to talk to us, and that distracted us for a bit. Tim made me laugh when he said, "Mom, I keep looking down the street with all the television trucks and rescue vehicles and thinking Eric's just gonna come walking down the street, look at me, and say, 'What the heck is all this? Man, you didn't even give me three days to get out on my own before you called for backup?'" Tim chuckled as he said that, but at the same time, a tear rolled down his cheek.

Rescue Command Center and the rock wall where Karen and Tim sat daily waiting for updates.

Most of the time we felt optimistic, but doubt did creep in now and then. Those brief moments were miserable. To stay positive, Tim and I talked mostly about what we thought Eric might be doing and how resourceful he was. We simply couldn't understand what could be keeping him from returning.

Mt. Baldy is a quaint little mountain village, home to around 475 residents. It has stone walls that tastefully border people's homes, adding to the uniqueness of this hidden treasure. There's a church, a lodge where people gather for meals, and a fire station that doubles as the command center for rescues. The fire station is attached to the post office, where people would pass by daily and check in with us. Mt. Baldy is a place where everyone knows everyone else, and I learned through their stories that the members of the community are always there for one another. When a flood caused a mudslide down the mountain into people's homes, neighbors and friends responded within minutes to make sure everyone was okay and to help dig out homes from the mud. It's the kind of place that if you mention to your neighbor that you like the motion-sensored floodlights on their home, those same lights will magically appear on your own home before you even asked to have them installed. The people and the structural charm of the place make this community unlike any other.

One of many stone walls that dot the village roads of Mt. Baldy.

Cell phone reception was almost nonexistent in Mt. Baldy, so I drove down the mountain to make any necessary calls. This day when I returned, Tim said, "Mom, Marcia Campbell stopped by and said that we're welcome to stay at her house if we don't want to leave at night. It's down that road to the left—the second to the last house on the right. She said no matter what time it is, the door will be open for us. She's great! You have to meet her."

As people would stop at the post office to pick up their mail, they would introduce themselves and let us know they were hoping and praying for the safe return of Eric and Gabrielle. I was able to meet Marcia the next day, and she was every bit as genuine as Tim had described her. Linda Frost checked in with us daily and brought us blankets for when the sun went down and it got cold again. These people treated us as though they had known us all their lives. The sincerity of their concern for Eric and Gabrielle was evident. Every time one of them stopped to check in with us throughout the day, it lifted our spirits. We had found new lifelong friends, and, as Linda said, family.

Marcia Campbell (left) and Linda Frost (right).

Local affiliates for ABC, NBC, and CBS were there daily, and they had been told that we preferred not to talk to the media. It seemed wrong to us, and most of the reporters respected that. One repeatedly asked if she could interview us. She was very sweet, and I enjoyed talking to her, but I told her I would not talk with her on camera. She didn't understand my resistance to being interviewed. I tried to respectfully explain to her that the focus needed to stay on Eric and Gabrielle. Being in a position of having to speak with emotion was also something we wanted to avoid. We were trying to keep our emotions under control, and continually being asked for an interview was making us uncomfortable. I felt guilty every time this particular reporter was around, waiting for the opportunity to ask one more time for an interview. Was I being uncaring and unappreciative of the reporters' efforts? It certainly felt as if I was. I know of no mom who wouldn't jump at the chance to talk about her children, but this wasn't the right time. No matter what I did or didn't do, I felt awful. Part of me felt sorry for this reporter. She said that if she didn't get a live interview, her boss probably wouldn't let her come back. It must be difficult to be a reporter in these situations and be told to "get the story." For us, we were glad the media crews were nearby to keep the story alive. We didn't want concern to dwindle and our hope for continued assistance to vanish.

Another reporter took a different approach. Tim and I didn't know for the longest time that he was even with a news crew. He told us he lived in the area, saw these rescues all the time, and that we were in good hands. His words made us feel better. He checked in with us every day to reassure us that everything would be okay. When I finally realized he was a reporter, I asked him why he wasn't continually asking us for an interview like some of the others did. He said he wanted to respect the fact that we had requested no interviews but that he was there for us if we ever needed to get a message across. I began to think of him as an ally, someone who truly cared about how this was going to end.

As we continued to sit on the rock wall, Tim showed me a trail map he had gotten somewhere. He located the trail Eric had taken. Tim boxed out an area where he thought Eric and Gabrielle could possibly be and narrowed that down by X-ing out the areas that increase in elevation. It would be kind of hard to fall up a mountain, he reasoned.

Frustration levels were rising as the day went on. We didn't feel our input was what those in the command post wanted to hear, so we tried to stay out of the way so they could do their jobs. That was very hard. When Tim had the opportunity to point out the trail, we were told that they could be anywhere, that people go missing all the time, take a wrong turn, get lost, and then show up hours later. Oh, my gosh! Is that what they thought happened to Eric and Gabrielle? Is that why I felt they weren't as concerned as I thought they should be? Or was I being overly sensitive because this was my son, and I wanted the rescue workers to feel the same level of urgency that I felt? That comment jolted me and added to our frustration. It made us feel that those in command thought Eric and Gabrielle were new to hiking and that they had just gone out for a leisurely hike and had taken a wrong turn. We knew that was far from the truth.

I hadn't shared with the search and rescue team anything about Eric's skill level and experience. Why not? Mainly due to the fact that Eric didn't share with everyone his military experience because sometimes that can be perceived as boasting. Given that, I was pretty sure he wouldn't want his proud mom to do that either. But things had changed, and I thought it would be important for those who were in charge of search and rescue to know that Eric wasn't wandering aimlessly without a clue as to where he was going. So I gave them some background information.

Basically, my message was like this: "Eric knows how to navigate. He knows how to solve problems. When he was in the military and did land navigation drills, he was the first to return, not the last. He's strong, resourceful, intelligent, and determined, and there is a reason why he hasn't returned yet. Something is keeping him from getting to us, and I can guarantee that it's not because he got turned around out there. He's a medic, so if one of them is injured, he knows what to do. He has climbed Half Dome, been in ice caves, and gone down moulins in Alaska, for starters. He would crawl on his hands and knees to get here if he had to, but he has to be making progress. There is some reason why he is not here yet, but being lost is definitely not part of the equation."

I was told that others had said Eric wouldn't split up from Gabrielle out there, and Randy, the sheriff's department liaison, asked Tim and me if we felt that was the case. I think they wanted to know whether they should be looking for two sets of footprints or one. We told them no, they would split up if needed. We felt that Eric would do whatever was necessary to survive. If Gabrielle were hurt and the only way to save her was for him to leave her, he would bandage her up as best he could, leave her whatever he could for survival, and build her confidence so she would know she would be okay alone until he returned. If he were hurt and couldn't move, I think the decision to separate would depend on the terrain and whether he thought Gabrielle could navigate her way out alone.

As Tim and I sat on our rock wall, I began thinking about Eric's younger years. I think moms always like to remember incidents that made them smile. I did a lot of reminiscing during those few days. I remembered how Eric's elementary teachers and the school secretary would send home notes or tell me something genuinely kind that Eric had done that made a difference for someone. Those reports were always followed with, "I just wanted you to know. Please don't tell Eric that I told you. You know how he prefers to keep those things to himself." I wish that I had kept some of those notes in a box somewhere.

I loved thinking about the things that made Eric happy. He started playing basketball when he was in third grade and loved every minute of it. That's where I started to see his leadership skills in action. And I enjoyed thinking about how Eric had always questioned and analyzed everything, starting at an early age. When he was three years old, the lock jammed on the door one time while he was inside the bathroom. It took quite a while to get him out of there, and while I waited for his dad to get home to help, I put my face near the floor and talked to Eric through the crack at the bottom of the door. After a while he asked me, "Mom, what are you wearing?" He was only three years old, and he was trying to gauge how much time he'd been locked in the bathroom. Had we all gone to bed, woken up, and gotten ready for another day? He was always thinking and analyzing.

My thoughts were interrupted when someone from the sheriff's department reminded us to get something to eat. I knew Tim had to be having many of the same thoughts I was having. At that moment, I could see the worry apparent on his face. With eyes watering, Tim said, "I can't. I'll eat when Eric eats." I felt the same way. In addition to suffering through the miserably cold nights, Eric and Gabrielle had to be getting hungry and weak.

The area near Mt. Baldy gets bitterly cold quickly as the sun sets. We waited for our daily briefing to find out what the teams had learned during the day. It helped to know what ground had been covered, but nothing could take away the emptiness that we felt when we drove off the mountain, knowing we were leaving Eric out there in the darkness and raw weather. Once again, Eric was going to tremble in the miserable cold with nothing to warm him. Tim and I barely said a word as we traveled back to the hotel.

Tim was frustrated because no one had been listening to what he had to say. I wanted him to be heard but also knew we needed to tread lightly. I didn't want our frustration to come across as though we were questioning anyone's professional decisions. That wasn't the case. What Tim was saying had validity, and it would have helped to have had that acknowledged. And here we were at the end of the day with no answers, while Eric and Gabrielle would spend their third night in the elements, the two of them alone in the dark with the wind whipping around them. The only thought that helped ease my guilty conscience for going to a warm hotel was that Eric would have wanted it that way. He would have told us that staying out in the cold and not sleeping would have helped no one. That is just a simple fact, and Eric puts a high priority on facts and common sense.

Day Four – Tuesday, April 9, 2019

Eric: In the Canyon

We had decided to descend into the canyon and get to the foothills, so we set out down the canyon shortly after sunrise. We wanted to give ourselves as much time as possible to get through whatever was beyond the 12-foot drop awaiting us. Thinking back to the first night, I remembered we had slept next to a dead tree branch. I figured when we got back to that spot, we could drag it to the drop and use it to assist us in safely climbing past that obstacle. As we descended the canyon, we routinely checked the compass, and it was consistently telling us we were heading south. When we got to the Day One campsite, we checked the direction. Our theory was confirmed. On Day One we had inconveniently stopped at the one location where the canyon twisted 180 degrees and the water flowed north for a short distance.

We reached for the long log we had slept next to on Day One. The log was quite heavy, so we decided to walk down to the drop and see if we could find a better log closer. I found one potential candidate and put it over the ledge, but that log ended up being a couple of feet short. At that point, I noticed a rock at the top of the drop that was wedged against the right wall of the canyon. I thought tying together a few of our jackets and securing them at the top by wedging a knot between the rock and the canyon wall might be easier than using a log. I tied three jackets together and used them as a hand line to control my descent down the ledge. After I had reached the bottom of the drop, Gabrielle unsecured the jackets and threw them down to me along with our packs. I leaned the log against the ledge for her to use as a first step. This allowed her to step down a little lower where I could safely lift her the rest of the way down. It was a euphoric moment. We thought we had overcome the obstacle that had prevented us from going home and now we would run out of the canyon. Little did we know that our challenges were far from over.

The first obstacle of many to come.

Shortly after getting beyond the 12-foot ledge, we came to a waterslide. It was a section about 15 feet long where the canyon came to a choke point and we would have to sit down and ride the watercourse down into a pool just like going down a waterslide. I turned to Gabrielle and confirmed that we were continuing to descend. She was worried because she didn't know if she could swim. I told her we could probably stand up in the pool and that if she was struggling, I'd jump in and grab her. We agreed to continue to push forward. I went down the slide, and it was actually a smooth, fun ride. The pool was about thigh deep on me, and the water was frigid. The cold water was rejuvenating for my tired legs, but getting socks wet is never a good thing. Gabrielle went down the slide next, but when she landed in the pool, she fell forward and was completely submerged. With her being so wet and cold, we knew we had to get home fast.

Water slide where Gabrielle was completely submerged.

We continued to push downstream and almost immediately encountered what appeared to be a 100-foot waterfall. I couldn't gauge the depth of the pool at the bottom. At this point, I figured the probability of death was pretty high. The entire time we were stuck out there, I was silently assessing our likelihood of death. Most of the time I had figured the chance of death was less than 20 percent. But when I saw a 100-foot drop in our path, my mind started spinning as I tried to think of potential solutions. My immediate impulse was to wait at the top of the waterfall and hope a rescue helicopter would find us. But I knew it would be hard for the search-and-rescue team to see us if we were deep in a narrow part of the canyon, especially since they hadn't spotted us when we were in a wide-open part of the canyon with a fire burning.

What options did we have? If we stayed where we were and no one came to get us and we were in dire straits, I would jump the waterfall, hope the water was deep enough and I would survive, and then I would exit the canyon and tell the rescuers where Gabrielle was located. That seemed like a bad option.

Could we try to ascend? No. The two obstacles we had just overcome would prevent us from ascending. The 12-foot drop meant we would have to find a way to climb back up to the top of that drop, but we couldn't even get to that obstacle because the current in the waterslide was too strong, preventing us from walking back up. Gabrielle was wet and cold and might succumb to hypothermia overnight. I mentally assessed that our chance of death at this point was at least 80 percent. Then I looked left, and it appeared that we might be able to climb to higher ground and start a big fire where the rescue crews might see us. We had been hearing helicopters all day long. Getting the attention of those helicopters seemed to be our best option. Outside of a little indoor rock climbing, neither of us had any rock-climbing experience. The route was steep but not vertical. I asked Gabrielle if she thought she could climb it, and she said yes. We wanted to get to higher ground as quickly as possible to signal for help.

I started climbing first and I discovered that a lot of the rock was loose. There were not many good holds. Eventually, the route got too sketchy for my comfort level, but I spotted a solid rock to my right. I traversed over and continued to ascend as quickly as possible to assess our situation ahead. I came to a ledge that provided an open view of the canyon where I could see a possible downclimb that would allow us to skirt around the 100-foot waterfall. It looked like we had some options again.

After taking a mental note of what was ahead, I went back to check on Gabrielle's progress. When I looked down, I found her on the left side, where I had started, hanging onto two branches coming from the left side of the mountain. She couldn't get any solid footing, and she was at risk of falling back down to the top of the waterfall. The rocks were too loose, and she asked how I had climbed to the top. I told her I hadn't taken the route she was currently on because it was too difficult; instead, I had moved to the right. She pulled herself up and found enough holds to get over to the right where she could walk up to the ledge. I was impressed with the strength she mustered to make that climb. Our bodies were getting worn down and the path was difficult, but we made it.

With both of us on solid ground again, we discussed our options. We agreed that the climb had been exhausting, and we didn't see much value in trying to climb higher. We still had a good amount of sunlight left at our location. We decided it would be best to get into dry clothing so our wet clothes could dry. We would try to signal for help during the rest of the day while we recovered. Then we would assess whether the downclimb or the 100-foot waterfall jump would be the best option for us if we weren't found.

We had changed into dry clothes and were setting up camp when we heard a helicopter and spotted it in the distance. We signaled it using a signal mirror, and it flew over us. We continued to signal to the helicopter with the mirror while waving my yellow jacket and the emergency blanket. We could see that the helicopter had "LA County Fire" on its side. The helicopter circled our position three times and came down to our elevation. On the second and third times circling us, someone stuck their head out of the helicopter. I took this as proof that he had seen us. We didn't waste any opportunity, though, because we had thought the helicopter on Day Two had seen us. We continued to signal the entire time the helicopter was in sight. The helicopter never signaled that they saw us, and it flew off. We were confident they had seen us and that they were going to bring a rescue team. Our spirits were high and we figured we would be going home in about an hour when the rescue team showed up. But after our Day Two experience, we decided to continue to prepare the campsite as if we would be staying.

The ledge that we were on was relatively small. We had to stay away from falling rocks from above and also avoid falling off the ledge. One spot had a large rock to protect us from falling rocks above and a smaller rock to prevent us from rolling downhill. I tried to use the ice axes to clear a bed, but the area was very rocky. The bed was simply going to be uncomfortable. I made a fire pit and Gabrielle collected firewood. As we prepared the campsite, we noticed some dark clouds at the mountain peaks. At first we couldn't tell if they were clouds or smoke from a fire, but it soon became clear they were clouds. As the clouds continued to descend, we became concerned about the possibility of rain. Nightfall was near and the rescue team had not shown up. We were extremely discouraged. We had been sure the rescuers in the helicopter had seen us, and we didn't know what more we could have done to get their attention.

We started a fire to prepare for the coming cold and darkness, but the fire didn't last long. The winds quickly picked up and blew embers all over, forcing us to put out the fire. Our night was going to be miserable. We went down to the bed area to try to be protected from the wind as much as possible and cover up with the emergency blanket. By this time the emergency blanket had been torn several times. We had enough pieces to somewhat cover our torsos. We used most of the other small pieces as insulation between us and the ground. We put our feet inside my long-sleeved shirt and put our legs inside the poncho from the survival kit just to get another layer between our bodies and the cold air. None of this helped much, as the night was miserably cold and windy. The wind tore our emergency blanket apart even more. The bed was a tiny space, and both of us kept having to shift around, as the rocks were putting painful pressure on various body parts. Several times throughout the night Gabrielle would say she wasn't sure if she could take it anymore. I kept telling her to stay strong, that we wouldn't die that night, and that we would get a hot shower the next day when we got home. Time was crawling by, making for the longest night of my life. It seemed to last forever. We didn't sleep for one second.

Karen and Tim: At the Rescue Command Center

This morning I went outside to the car and ran into Eric's dad, Mike. He had arrived late the previous night with his brothers, Don and Steve. Mike was visibly upset as he told me, "I can't go through this again. You know what today is." I responded, "Yes, and I need to not think about that. Everyone else needs us right now: Eric and Gabrielle and Tim and Nikki. It's important we focus on today."

It was on this day in 1985 that we had held our daughter Tara for the last time. She had survived almost a year with leukemia and then lost her battle shortly before her third birthday. Yes, I did remember what day it was. It sounds silly, but if our end was not a good one on this venture, I really hoped it wouldn't be today. Of course the date didn't really matter. A day is a day is a day. Instead of engaging in a conversation, I simply hugged Mike. I prayed for God's peace to help us stand united while we found Eric.

We arrived at the fire station around 7:30 a.m. and saw longer lines of television trucks and rescue vehicles. The increased level of activity with the additional rescue workers brought us renewed hope. The temperature started to warm up as it got later in the morning, and that always took some pressure off because lack of warmth was one less thing to worry about. The sun was at least warming Eric and Gabrielle, wherever they were.

Tim took Mike, Don, and Steve up to the saddle because they wanted to see where Eric had last been. While they hiked, I drove to town to take care of Eric's cat, DJ. By now the apartment complex office had talked to the powers that be, and they had decided it would be okay for me to feed Eric's cat. Poor DJ really loves only one person in the world, and that's Eric. DJ was lost and confused as he hid under the bed. I had to smile when I looked around the apartment. Most things were in place, neat and tidy like they always were. At the age of three, Eric had already started refolding and reorganizing his laundry after I had done it, so he has had a lot of practice in applying his organizational skills. And let me just say that my laundry skills aren't too shabby. Eric just wanted the folding to pass *his* inspection, so clearly he saw room for improvement when *I* folded the laundry. He wanted the short-sleeved shirts to be in one pile and the long-sleeved shirts in another. So when I saw that his bed was unmade, I whispered to DJ as I stroked him, "He'll be coming back, DJ. He has to because he forgot to make his bed."

While driving to Mt. Baldy, I received a text.

"Hi, Karen. This is Ryan, Eric's buddy and coworker. I just wanted to let you know both Gabby and Eric have been on my mind since late Saturday. They're both great people and I've been pretty torn up. Wishing and hoping for the best still. I hear there's some good leads today and wanted to offer any assistance I can to you and Tim. I have a spare room for you two, but I'm about an hour away from the mountain. Let me know how I can help."

I responded, "Thank you so much. If you or anyone at the office would like to come to the rescue center when you're done working, we would love to see you. We hiked the route yesterday, but they prefer we stay at the command center. One minute of waiting seems like hours, and the conversation would be welcome. Truly no obligation."

"We will take you up on that," he responded. "Eric is strong, determined, and intelligent, one of the smartest guys I know. I really believe he's out there leaving us breadcrumbs."

This was my introduction to what would become our lifeline for the next two days.

I returned to the command post and took up residence again on our rock wall. Tim was talking to Mike, Steve, and Don. This gave me an opportunity to sit quietly with my own thoughts again. When people would walk away and I was by myself, I could whisper, "Eric, I love you" while hoping God would take those words and put them in his ear. Back came the thoughts of him lying on rocks injured but alive. I just wanted him to hear me and know that I loved him and was so very proud of him. There were so many conversations I had wanted to have, so many things left unsaid. That left me feeling empty, so I turned to the happier times.

I laughed inside when I thought, "You have to come back, Eric. You're the only one who listens to my political views and agrees with me. Or perhaps that silent nod you give every now and then is just a way to avoid a discussion you don't want to get into at the moment. That's okay. Faking it every now and then is never a bad thing. But it all comes down to values, and it's always nice to have someone in the same corner. I wish I would have told you that meant a lot to me. For now, I'll just whisper, 'I love you,' and hope the sound travels to your ear."

I was surprised that the rock wall didn't topple from us sitting on it with our roller coaster ride of emotions. It's where I sat as I tried to pray for Eric and Gabrielle's safe return, and all I could ever mutter was, "God, please keep them safe." Prayer was different for me now than it had been when Tara was sick. Back then, I didn't truly believe God would hear me. I had made my share of mistakes in life, and I didn't think I was worthy to have God listen. I prayed every time Tara had a bone marrow procedure. They were so painful and her worst days. I wanted me to feel the pain instead of her. Those prayers were never answered, so in my faithless mind, I figured God wouldn't hear me if I prayed for her to survive.

Many years had passed since Tara died, with lots of life experiences where I've needed to lean on Him daily. I've learned how wrong I was. Today I could pray for God to keep Eric and Gabrielle safe and know that I was being heard. If the kids could come out alive, even if they were severely injured, I knew Eric could do the rest of the work to heal. He showed determination in everything he did, far beyond what I've ever done. If Eric wasn't coming out alive, I knew God would help us get through that too. For now, I simply asked Him to keep them safe. Then my thoughts wandered elsewhere. I knew for certain that my friends back home were praying. I could feel their support.

Eric's coworkers came after work to pass the time with us. I enjoyed putting faces with names that I had heard on the phone when talking with Eric. We introduced ourselves to one another, and I could see in their faces how troubled they were that Eric and Gabrielle were missing. Nicole had tears in her eyes as I met her the first time.

As they began to talk, I noticed that their personalities were exactly as I had envisioned. I'll never forget Ryan starting off the conversation with his one-word description of Eric: "Eric is very … (loooong pause) …direct." That comment caused everyone to laugh. Yep, you always know where you stand with Eric.

And then the stories began. The stories of Eric buying a new jacket, sliiiiding into Ryan's office, and saying, "What do ya think? Looks pretty fresh, eh?" as Eric moved his hand down the side of his jacket.

We shared pictures, laughed, and told stories for the longest time. I had to swear them to secrecy on some, like the one of Eric and Tim on the carousel when they were 9 and 13 and then the same shot 20 years later on our return trip to Disneyland a mere three months ago. They appeared to have more fun the second time around as adults. And then there was the magic shot with Eric holding Tinkerbell in his hands. They shared pictures of Gabrielle and Eric hiking together.

We talked about Eric's sense of humor and how he made us laugh—like the time when I was driving and Eric had stated out of the blue, "I *feel* like you drive too fast for your skill level."

I let him know that his comment might be perceived as an underhanded insult. He said, "Oh, no, Mom, that was pretty direct. Just stating a fact."

A few of the happy memories we reflected on. A good distraction.

DAY FOUR — TUESDAY, APRIL 9, 2019 47

My response: "*I feel* it might be a very long walk home."

It made me laugh to remember that drive, but it also made me miss him even more. I wanted to see him smile and laugh again.

I don't know if his friends realized how much that visit meant to us. It was hours of nothing but happy thoughts and laughter. No worries. And as I listened to each of his friends and coworkers talk, I saw how much they truly knew Eric as a person and appreciated him. It was easy to see that Eric and Gabrielle were not only their coworkers but also their friends whom they truly cared about. Those coworkers and friends had taken the time to come, sit with us, and relieve some of our stress—stress we didn't even realize was as heavy as it was …until it once again got dark and cold, and his friends left to go home. The clouds suddenly got dark and sank until they appeared to smother the mountaintops. The temperature dropped quickly. This brought us back to our reality.

We had gone from laughing and feeling so alive to being engulfed by darkness and fear. It was like the panic that I felt after Tara's funeral. We had an abundance of joy as we shared stories with one another, and then people started to leave for home. I wanted to block the doors and say, "Wait, we can't be done yet." It was a fear of finality and a fear of being left alone with an indescribable emptiness. It was exactly the same feeling I'd had on the mountain. I was afraid I was not going to be able to hug Eric again and tell him I love him. I wanted to tell his friends, "Please don't leave." As long as they were there, Eric was alive. I wanted them to stay so we didn't have to feel this way. But they were gone, and it was dark, and the wind was again piercing, and Eric and Gabrielle were still out there alone.

I had been so hopeful that today would be the day when they were found. Randy, the liaison from the sheriff's department, gathered us together outside the command post to report that they had no news for the day. The confirmation numbers of rescue personnel lined up for the next day seemed fewer than what we'd had today. I couldn't speak. I had nothing left in me. I simply turned to Tim, buried my head in his chest, and sobbed. The fear that I had been trying so hard to hide came out in one single moment. I wanted to hold myself together. I didn't want to be the needy one, but I had to lean on Tim for a little while. I said, "I love you" repeatedly to him. If Eric couldn't hear my words, I wasn't going to have another son go one more minute without knowing how much I love him. I finally lifted my head, and no one was left but Tim and me. He said we should probably go to the car and head back down the mountain, and he was right. We didn't say a word on our trip back to the hotel. Here we were, deserting Eric and Gabrielle again, while they had to find a way to endure another frigid, dark night.

Willie descending waterfall.

Day Five – Wednesday, April 10, 2019

Eric: In the Canyon

We anxiously waited for the sun to rise and bring us some relief from the unbearable toll another cold night had taken on our bodies. We were shivering so violently that it would be impossible to make a technical downclimb until the sun had warmed us up. Because we were on the east side of the canyon, it was late into the morning before we were warm enough to attempt the climb. As we tried to forecast what the downclimb would entail, we could see that it looked possible. Again, neither of us are climbers, and downclimbing is harder than regular climbing, but it seemed like this option would give us better control of the outcome than jumping 100 feet into a pool of unknown depth. After two near-rescues, we had lost all faith in search and rescue. We also didn't have much interest in staying at our location after the previous night's experience. Our only option was to walk out the south end of the canyon.

To give our bodies a little more strength for the downclimb, we each ate half of our second-to-last energy bar. Then we proceeded at a gradual and measured pace, checking every hold before committing to it. The first pitch was a relatively easy descending traverse. It was followed by a couple more of increasing difficulty. Then we came to a 30-foot vertical crack. We were able to wedge ourselves into the crack and safely descend to a sloped ledge of loose rock. We descended on the ledge for a bit and came to a dropoff on the right, but we saw that we could descend farther left and come to a reasonably sloped bed of loose rock. After descending that slope, we ended up at the watercourse. At this point, we felt confident. We had successfully maneuvered around what had appeared to be an impossible obstacle when we were stuck between the waterslide and the waterfall. We just didn't know how many other obstacles we would have to overcome along the way.

At the watercourse, we rehydrated to prepare for the rest of the descent. We immediately came to a rock we had to slide down before making an 8-foot drop to the riverbed. This appeared to be a simple task. As I started to slide, I moved much faster than expected and hit the ground harder than I thought I would, which caused me to fall forward. I caught myself with both hands to prevent my face from smashing into the rocks, but my right index and middle fingers were jammed and my left palm was bruised. I fell flat in the water, soaking the front of my clothes. Gabrielle made a much more controlled descent down the rock and landed on her feet.

Eric had attempted to slide down the rock on the left when he injured his hand

We continued down the canyon until we encountered another obstacle—a waterfall that was 30 feet high. I initially wanted to stay in the watercourse and thought we could survive this jump. Then Gabrielle noticed what appeared to be a path going up to the left. Maybe we could skirt around this waterfall too. We decided to check it out and started climbing the path. There were some footprints on it, and we saw two yellow pieces of webbing tied to a tree. A red piece of webbing was tied to a different tree at the top of the waterfall. This was a good sign in our minds, as this was the first time that we had seen any signs of human activity. We thought that maybe an established trail was nearby that would allow people to access this area from the bottom of the canyon. We continued to climb the path to see if it would easily allow us to avoid the waterfall. But then the climbing became insecure and didn't bring any promising leads.

As far as we could see from our vantage point, the watercourse didn't have any significant obstacles. We didn't want to climb too high and spend another cold night atop a ridge and then face a difficult downclimb, so we decided that getting back down to the canyon floor was our best option. I retrieved the red webbing from the top of the waterfall, then tied the red webbing to the two yellow pieces of webbing and five of our jackets and secured the makeshift rope to a tree about 40 feet from the canyon floor. We couldn't see the canyon floor from our location, so we didn't know if our rope reached the ground.

Because we wouldn't be able to see each other and we knew that the crashing water would prevent us from hearing each other clearly, I worked out simple commands with Gabrielle. When I made it to the canyon floor I would yell, "I'm down." She would then respond with "Confirmed" or "Say again." I would then determine whether our rope reached the ground and could be used to lower the packs or if Gabrielle should just throw the packs down. I would yell, "Pull up" or "Throw down."

Handline made of jackets.

Close-up of Eric's handline knots.

I then showed Gabrielle how to wrap the rope around her arm to secure her hold and slowly descend, since we didn't have a harness. When I was descending, the wall got slippery when I was about 15 feet above the canyon floor. My feet skidded and my body slammed into the mountain, but I held onto our rope of jackets. After catching my breath, I lowered myself to the canyon floor and signaled to have the packs lowered on the rope. Gabrielle lowered both packs on the rope as planned.

Gabrielle began her descent and was doing well until she hit the same spot where I had lost my footing. She also slipped but couldn't hang onto the rope, and she fell about 15 feet to the canyon floor. I helped to brace her fall, but falling that far stung her already worn-out legs. She said she was okay to continue on. I tried to untie the bottom jacket—Gabrielle's jacket—since it was the only one I could reach, but the knot was too tight. I pulled out my knife and cut the sleeve of my shirt, the second to last in the rope, so we could at least keep Gabrielle's jacket. Now I only had one T-shirt and Gabrielle had a T-shirt and two jackets. We thought it would be important to get out of the canyon before nightfall because we wouldn't have enough clothing to keep ourselves warm through the night.

We turned to walk downstream and made it beyond what I could see from atop the waterfall. We identified an open, sandy area where a lot of dead wood was available, making it a good place to build a fire and bed down if we got stuck. About 20 feet later, we encountered a 15-foot drop. We knew we could jump it, but we didn't want to lose the excellent sleeping location in the sandy area we had just passed. Attempting to sleep on rocks had taken a toll on our bodies, and sand was a welcome sight. We decided to start a fire and postpone the jump until the next morning.

Eric at the bottom of rappel. The waterfall was averted by creating handline from jackets.

When we got back to the campsite, Gabrielle gave me one of her remaining jackets. It was way too small to fit me, but at least we both had a shirt and a light jacket. We gathered firewood and got the fire going right away. We did what had become our typical nightly preparation by getting out our chem light, flashlight, and ice axes for defense. We treated our injuries, applying topical antiseptics to cuts and taking pain killers. After doing that, we tried to assess how to best use our available resources to keep warm. The emergency blanket had been decimated at this point, with only two significant pieces left. I wrapped one of them around Gabrielle's torso and taped it down. She then did the same for me. We would use the rest of the scraps as an insulator between our bodies and the sand. We planned to put our feet inside my waterproof pack cover and both our torsos inside the poncho. We were set for the night and were huddled around the fire.

Near dusk we heard a helicopter downstream from us. It sounded like it was in the canyon. We looked over to see a helicopter hovering just downstream of us. We immediately popped up. Gabrielle grabbed a piece of the emergency blanket and waved it over her head. I reached for the chem light, cracked it, and started spinning it. The helicopter shined lights on us and activated a short siren noise (bleep, bleep). We thought we could make out the words "County of San Bernardino" on the side of the helicopter. The pilot then spoke to us through a speakerphone. He told us to come out to the open so we could talk. We ran over to the ledge. The pilot confirmed who we were and asked if we were injured. We thought the pilot had asked if we were able to move to a location where he could set down his helicopter to pick us up. It sounded like he wanted us to move about 50 meters in a certain direction, but we couldn't be sure because the noise from the helicopter was bouncing off the canyon walls and making it difficult for us to hear what he said. He told us to wait where we were and that he would be back in one hour with a different helicopter.

We couldn't have been happier or more relieved to know that in one more hour this ordeal was finally going to be over. We waited at our fire and packed our gear so we would be ready to go when the helicopter came back. We also split the final energy bar and ate it as a celebration.

About an hour later, a helicopter appeared in the canyon downstream of our position. We flashed our flashlight and swung our chem light to make sure they saw us. The pilot then spotlighted us. We knew they had seen us, so we poured water on our fire and went back to the ledge. The helicopter flew upstream, and we chased it upstream. Then it flew back downstream, and we ran back downstream to the ledge. Then it flew away. We didn't know what had happened, but it was getting cold, so we went back to our campsite and once again started a fire.

After about 10 minutes, we heard someone yell from downstream. We yelled back to acknowledge him. It was a member of the search and rescue crew. We extinguished the fire and ran back to the ledge. The crew member said, "The question is, how do we get you down here?" We spent about 10 minutes discussing our options for getting down to his level. We didn't feel comfortable with any of the available options, so I asked him if he had a rope.

His only rope was a 12-foot-long piece of webbing. I tied it around an anchor and put it over the ledge, but it wasn't long enough to let us safely descend. We had two options: He could get a longer rope or insert the helicopter on top of our position. He said the helicopter couldn't safely insert directly above our campsite because the canyon was too narrow, so he was going back to the helicopter to insert upstream of our campsite. I told him not to go too far or else he'd be upstream of the 30-foot waterfall.

Gabrielle and I walked back to our campsite to start a fire for the third time. A while later a helicopter appeared directly above us. It was blowing small tree branches down on us, and embers from the fire were blowing everywhere. We stood up and looked downward to protect our faces from debris coming from above. I felt something hit the back of my neck, and Gabrielle felt something hit her side. I looked down to see a large log next to our feet. The rescuer must have broken a large portion of the tree as he was descending, and the log hit us. The rescuer disconnected from the cable, and then the helicopter flew away. The wind from the helicopter had extinguished our campfire. The three of us chased around the embers so we could put them all out.

The rescuer started putting a harness on Gabrielle. He said she needed to take off her pack and that he would bring it up later. I then removed my pack. He looked at my pack and asked what I had strapped to the outside. I told him there were two ice axes and trekking poles. He said they might not be able to come with us. I asked, "They have to stay here?" He said, "Maybe." I then put my pack next to Gabrielle's. The helicopter returned and hoisted up Gabrielle. The rescuer then put a harness on me and I was hoisted up. I looked around as much as I could on the way up to the helicopter. I wanted to get an idea of my surroundings and see how far we had come and how much farther we would have had to hike to make it out. It was too dark to get much of a view.

When we were hoisted up, we were not allowed to touch the helicopter. A crew member rotated us inside and clipped us in. A fourth member of the rescue team watched out the window and communicated with the pilot, as our blades were just 10 feet from the canyon wall. It was quite impressive to see the pilot and the spotter working together to hover inside the canyon without hitting the walls. When the rescuer was hoisted up, we noticed that he didn't seem to have enough gear with him. Gabrielle asked if he had got all of our gear, and he said that he could only bring her pack and he had left mine. If I had known this might happen, I would have grabbed important items out of my pack and put them in my pockets. Oh well, at least we would be warm that night, and the saga was coming to an end.

We flew for a short distance, landed, and were handed off to the San Bernardino Sheriff's Department. They brought us to an ambulance for medical exams. While I was being examined, the officer told me, "Eric, your mom is here." I said no. "Your dad is here." No. "Your brother is here." No. He said, "No, they ARE here." I had thought he was asking me if they were here so I could have them come pick me up. I was going to tell him that all of my family was out of state and that I needed to call and tell them what had happened. He then told Gabrielle about all the people who were there for her. I turned to Gabrielle and joked, "Well, look at all the people that care about you." He told us that the fact that we were missing was big news, and I was shocked. I didn't know why anyone, outside of our friends and family, would consider our whereabouts to be newsworthy. After our medical exams, we were brought to a police trailer where we were debriefed. The officer interviewed us and we told him everything that had happened in a step-by-step recollection. After the debriefing, we were driven to the police command center where we were reunited with our families.

Last camp and extraction point, 04/10/19.

Eric's pack patiently awaiting his return.

Eric starting his rappel down the last waterfall Gabrielle and he would have needed to find a way around the next day, had they not been rescued.

Sonny rappelling down the last waterfall.

Karen and Tim: At the Rescue Command Center

After sleeping for about two hours, I awoke in a panic with my heart racing. I thought about the reduced number of volunteers we had been advised of the day before, and I was afraid people were going to start giving up on finding Eric and Gabrielle. I had to do *something*. I needed to thank those who had given their time to the search efforts thus far. I remembered the reporter who had told me, "I'm here if you ever need to get a message out." Well, I needed to do that now—I wanted to thank the people who had put so much time and effort into hiking up and down the mountains searching for Eric and Gabrielle. I wanted the volunteers to know how much their efforts meant to us. I had no idea how to do that or even if it was possible, but I was going to try. I texted the reporter and said I needed to get a message out. Then I woke up Tim and said we had to head back. Although he was completely exhausted, he jumped up and prepared to head out. As we left the hotel, I asked the manager for blankets so we could keep the kids warm when we found them. As they had done since we arrived, the staff once again accommodated us.

The search-and-rescue trucks continued arriving around 7:00 a.m. as we reached the command center. We always watched them load up all their equipment and their packs before they left. I desperately wanted to go over and hug every one of them and thank them in person. I wondered whether they knew how much we appreciated their efforts. How would they know unless we told them? But we also wanted to stay out of the way of operations, so I refrained from interrupting them. I spotted an orange truck that Tim and I had both noticed during the past few days with "Alpine Unit" written on it. We watched the crew members strap on their gear, which seemed endless. How in the heck could anybody walk with that much gear, let alone navigate the rugged terrain to which they were headed?

Tim walked over to me and discreetly pointed to them and said, "Have you been getting a load of these guys? These guys are beasts!"

"Yep," I replied. "You know what Eric would say if he saw them."

Tim laughed, "I know, right! 'Those guys have all the bad-ass gear. I wanna be on their team!'"

Starting the day sharing a laugh felt good. We later found out this was one of the crews that planned to stay in the canyon that night. They felt they weren't making enough progress each day when they kept starting from the command center, so they planned to spend the night in the mountains and start again the next day from a point deeper in the canyon.

When the reporter who had offered to help arrived, he got out the message as I had asked. This whole ordeal was about more than Eric and Gabrielle being missing and what they were going through. It was about those who had given their time to help people they didn't even know. It was about the people in the village who had stopped by and offered their support. It was about friends who sincerely cared and prayed every day for Eric and Gabrielle's safety. I have no idea how the news had spread so quickly, but we were grateful for the outpouring of support from everyone, including total strangers. There was really no way I could put my gratitude into words. I've tried to do so on many occasions, and I simply can't. But I hope people know that their support was valued. Encountering goodness in the world is always an empowering experience.

That day we talked to detectives from the San Bernardino County Sheriff's Department. One of them questioned Gabrielle's roommate and another talked with me. I can't remember all of the questions we were asked, but I got a sense that this gentleman was not happy. He reiterated how the rescue efforts were the best in the state and how hard they were working. I told him, "I don't think I've ever appeared to question that, and in fact, I have stated my support. Just last night in the command post I voiced my support for them as someone else was questioning procedure."

He said, "Well, it sounds like somebody has a problem with their efforts." He mentioned Tim's map.

I told him, "Yes, my son has a map and has boxed an area that he would like to be the focus. Neither he, nor I, understood the reason for inserting rescue crews on nearby mountain peaks if Cucamonga Peak was Eric and Gabrielle's destination. We are clueless when it comes to operations in an event like this, but Tim's heart was saying that his brother was closer to the boxed area. Frustration levels were getting high, and Tim apologized for his persistence. If that person isn't satisfied with the apology, then they need to speak up and say something directly. Tim can't fix what he doesn't know is broken. Even though he was highly stressed with this being HIS brother who is missing, Tim took the time to be accountable. If his apology was not adequately accepted, let me know. I'm sure he'll try again."

I added, "I think if someone were able to put themselves in our shoes, they would understand. That's what we've tried to do, which has helped us to respect the boundary lines. We've done our best to keep our emotions in check here, but we're human, and this is highly stressful. It's been an emotional roller coaster, and things aren't always going to go perfectly."

He seemed to accept what I said. I was glad, because otherwise we would have had a long, silent ride together. We were now headed to Eric's apartment in the detective's vehicle. They wanted to get some clothes for search dogs to use and to get Eric's computers. We noted the hiking packs and jackets still at Eric's apartment in order to narrow down and confirm what he was wearing and the pack he had actually used. The detective took pictures of Eric's hiking maps. We grabbed Eric's computer and the clothes and left.

When we returned to the rescue headquarters, Tim and I went back to our rock wall. Waiting, waiting, more thinking. Tim spent a lot of time with his dad and uncles. That created time for me to get lost in my own thoughts, and a good share of them involved bracing myself for what I thought was probably going to be our reality. In my mind, there was no way that Eric wouldn't have found his way out after so much time had passed. If he were injured, he would have found a way to crawl to an open area where he could be seen. I had to start picking up the pieces now for what I thought was more than likely going to be a sad outcome for us all, but I didn't want anyone to know what I was thinking. Everyone wants to hear a story about the family that never gave up hope, but that wasn't my perspective that afternoon. I had struggled to stay focused on the positive of WHEN they were found. Instead, it was now IF they were found. To some people, my thoughts may seem dark. However, I was trying to mentally prepare for the worst so when the time came, I would know how to best support my family.

I thought about whether I should bring Eric home to be buried by Tara. In my heart, that would have been selfish because it was me who would have felt better with them both physically being next to each other. Or should Eric be cremated and his ashes spread over a place that I know he loved and appreciated? It would be more fitting for him to be outdoors with the beauty of the mountains. Should I ask Nikki and Tim what they wanted to do? They had a right to help make that decision. But what if they had different opinions? Then one of them would have to live with not being at peace with where Eric was laid to rest. I couldn't do that either. I needed to make up my mind so that neither of them would have to think about that. There were so many questions to answer. Should I send Tim and Nikki home ahead of me and drive Eric's things home by myself? Should I ask Tim and Nikki to stay? Should we send someone back for Nikki's sons—Boomer, Dysen, and Renaud? How would we make sure Eric's nephews had closure? Who would care for Eric's cat, DJ? He was going to be lost without Eric. Was Gabrielle okay out there, or had she met the same fate? I wanted to know her. I had never met this friend with whom Eric had spent many weekends hiking. We knew from their coworkers that she was warm-hearted and quiet, but I wanted to know the person who had spent Eric's last days with him.

Then my mind drifted to thinking about my dad. God, how I wished my dad were still alive and sitting next to us. I missed him so much. There's no way he would have allowed us to go through this journey without him. I envisioned him sitting next to me on that rock wall wearing his khaki pants and white tennis shoes, with his legs dangling down next to mine, not quite reaching the pavement. That thought made me smile. He was a good man known for his compassion and generosity with his time. He was extremely proud of his grandchildren and enjoyed watching them grow and make their own path in the world. I'm glad my kids got to know him, because they each have a piece of him within them today. Dad always smiled when he asked what new project Eric had taken on. When I traveled with my parents and we stopped to see Eric, Dad wanted to see the basement that Eric was finishing. He was quite impressed and proud that every bit of the project had been done with Eric's two hands: the designing, adding a window, framing, drywall. Dad asked how he was able to get all that sheetrock in the basement, as the yard was on quite a steep slope. Eric just flexed his arms, and Dad responded with a smile. They are both quite alike in many ways, and their drive and ever-present work ethic is something that definitely matched. "Never pay someone to do something you can do yourself" was their motto.

Gone, yet still here.

As the sun was in full force and finally keeping us warm again, Randy, our liaison, came to tell us that we should get something to eat. His interruption was welcome. I wished time could move faster. Nikki was arriving tonight and I was anxious to have her with us. I was eager for all of us to be together in one spot again very soon. She handled everything so well, and it had to have been hard on her being the one who was holding everything together back home. Now it would just be a few hours before we'd all be together waiting for Eric to return while he was working to get to us. Oooooh, yay!!! Thinking with hope again was comforting—a bright spot after focusing on planning Eric's funeral.

During rescues, San Bernardino County provides a food truck staffed by inmates who are, believe me, the *best* cooks you could imagine. Every time Tim and I went to the truck for anything, the gentlemen staffing it always let us know they hoped for the best for Eric and Gabrielle. I sat and watched them as they interacted with others, and they were genuine with everyone. They didn't have to go out of their way to lend their support, but they did. Their circumstances were difficult, and yet they still chose to offer what they could to total strangers. I wish I would have told them how much I valued them during that time. I wanted to know their stories, but the time wasn't right for that conversation, and I also didn't want to seem intrusive. All I knew was that they had shown compassion to us over the past few days. One never knows when a simple act of kindness can mean the world to someone. This was one of those times. Someday I hope to thank them and let them know that I've hoped their second chance in life would turn out to be a good journey for them when that time comes.

Eric and Gabrielle's coworkers came up again that evening. Tim and I appreciated their company. I didn't remember the conversation in as much detail as I had the night before, but their presence made us feel better. Eric's uncles stood in the circle of Eric's friends sharing stories that kept us laughing. I was grateful they were there.

Shortly before the sun went down again, around 7:00, Eric's coworkers left. It was the darkest of hours. They left, the sun set, and again the weather turned wickedly cold. I felt the life being sucked out of me. How could Gabrielle and Eric possibly survive another night like this? I felt like they were still alive but certainly injured, and the odds of them enduring another cold night dwindled. I left most of the praying to my friends. For me, it was a continuation of asking God to please keep them safe.

I had gone to the side of the building with Gabrielle's roommate. It was pitch black there. Quiet. After we had talked for about 45 minutes, Tim came around the corner of the building and said with urgency, "Mom! Come on, Mom." He helped me get up off the cement step I was resting on. "They found something!"

We went to the front of the building and waited for Randy to update us. "They found two sets of tracks," Randy reported.

My heart leapt inside my chest. Oh, my God, could it be true? Was this a sign that they were coming home? I didn't have much clarity during those next few moments, but what I remember clearly is the moment Randy told us that the tracks were indeed theirs and confirmation had been made that Eric and Gabrielle were found. He told us the rescue crew needed to get another helicopter to get them out.

"Are you sure it's them?"

"Yes, it's them."

I repeated, "But I mean, are you REALLY sure?"

Again, Randy confirmed it was them. Tim and I hugged each other—this time with joy. Thank God this was over! Tim and Nikki would still have their brother, and our prayers to bring Eric and Gabrielle home safely had been answered. We were astounded and completely elated! I had never felt this kind of joy, happiness, and gratitude in my life. The rescue crews had never given up, and they had accomplished what had seemed to be impossible.

Although phone reception was spotty, my phone beeped. Eric and Gabrielle's boss, Ryan, Charles, and Nicole were calling to ask, "Is what I'm hearing true?" Thanks to the news crews that stayed that night, the news had spread quickly, much more quickly than we could get the word out because our text messages weren't reliably being transmitted. Tim and I tried to respond so everyone would know that Eric and Gabrielle were indeed alive! I couldn't get any response to my attempts to make contact. Tim, however, had gotten contact from Nikki. She was off the plane, and the news on the monitors in the airport said they were found. What it didn't say was whether they had been found alive, so Nikki was in limbo. When she was finally able to reach Tim, he reassured her, "Yep, they're found and they're fine. Eric was just out there roasting marshmallows," a comment Nikki took literally as she wondered why in the heck her brother had packed marshmallows to go hiking.

One might think the minutes would have felt like hours until the chopper finally picked up Eric and Gabrielle and brought them to the command post, but that wasn't the case. Time goes much faster when one feels at peace. While we were waiting, Nikki arrived. I remember her getting out of the car and running over to us. I held her for the longest time as she commented, "This is what true happiness feels like!" Indeed.

The news crews' camera lights came on as the ambulance arrived with Eric and Gabrielle. Rescue workers got them into the command post without much interruption. Search-and-rescue members created a diversion so Eric and Gabrielle could sneak inside the building through a side door. The rescue team wanted to talk to them about what had happened before everyone else bombarded them with questions, so we stayed outside until we were given the all-clear to go in.

Eric and Gabrielle both looked thin. When I hugged them, they seemed frail. They were a bit drawn in the face and very tired, but they were alive. I forewarned Eric that television cameras were outside. I let him know that many people had been incredibly gracious to us throughout the search, and I asked him to thank them for their efforts on his way out. For someone who had been missing for five days in the mountains, he did an incredible job of articulating how grateful he was that they were found that night. Then, when we got to the car, I had to tell him, "Well, just so you know, Eric, somehow you two made national news the past five days." I'll never forget his response: "Don't people go missing in the mountains every day?" Well, yes, correct …just not like that.

We went to the hotel. I couldn't understand how he had any energy and didn't ask to go home right away so he could just crawl into bed to sleep. I'm not sure he even sat down before he asked for a map. He wanted to confirm why his compass had showed they went the wrong direction the first day. He was fairly certain the watercourse had shifted, but he wanted verification.

"HAAAAA… THERE IT IS!" he exclaimed when he found the S turn in the canyon.

"Aha!" Eric locating where the watercourse shifted and flowed north, throwing off the compass reading.

I sat in a chair and watched the kids talk, smile, and laugh for the next few hours. Together again. How could anyone ask for anything more? Life felt pretty darn good. Although we have so much to look forward to and will have many more moments of joy as a family, I doubt we will ever reach a peak of happiness that's higher than the one we reached that evening.

Best Siblings Day Ever! Nikki, Eric, and Tim.

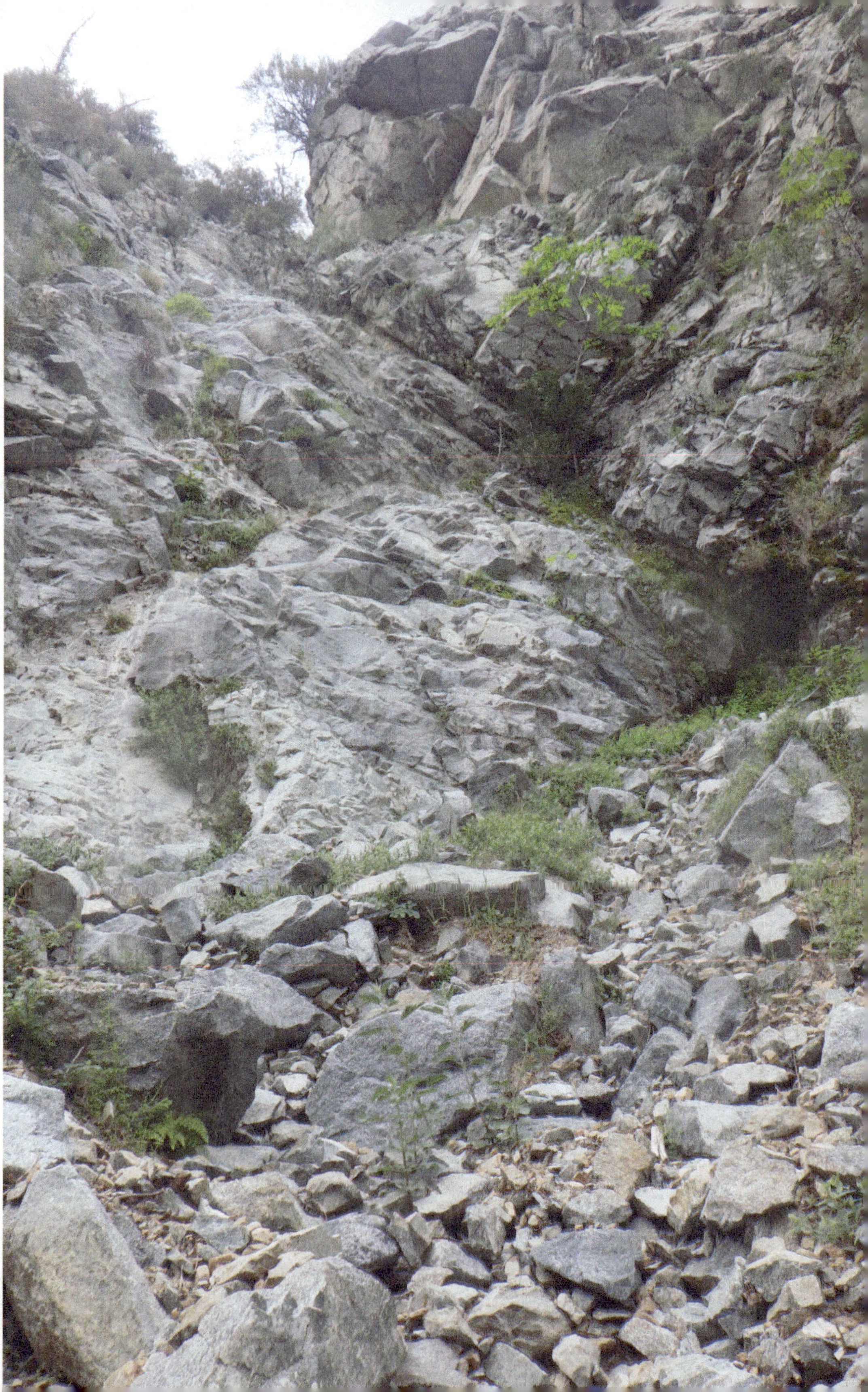

Thank You to All of the Search and Rescue Teams

Since Eric and Gabrielle's rescue in 2019, I've learned more about the search-and-rescue teams in California. In addition to holding down jobs, these people dedicate part of their free time to help with rescues. Their training is intensive and thorough, and the various teams have different requirements. To provide a basic idea of what a team member goes through, Art Fortini, Operations Leader, shared the training undergone by members of the Sierra Madre Search and Rescue Team.

The training takes 18 months, beginning with classroom sessions once a week. After a few weeks, the volunteers enter a period of extensive field training. Navigation is tested by requiring trainees to find preset control points using nothing but a compass. No GPS is allowed. Once the volunteers pass that test, they must pass a written test. Next they are taken to a secret location for two days and given an assignment that will assess their mental toughness in the field. After passing all three areas, they receive their blue patch and are granted full team membership. The unit then enrolls the members at a local community college where they take Emergency Medical Technician training. The search-and-rescue team provides funding for training and some gear; however, a team member can expect to contribute approximately $2,000 of their own money over the first year.

The search-and-rescue teams are not limited to searching in their own area. When a team needs additional help, it sends out a call to the other teams in the county requesting their assistance. Eric and Gabrielle's search was headed by West Valley Search and Rescue and assisted by Sierra Madre Search and Rescue and others. One factor that determines whether additional support is requested can be weather. If winds are high and helicopter support is limited, the teams compensate by putting more crews on the ground. The experience level of the missing hiker and the difficulty of the terrain also play into whether more support is needed.

As we watched team members from our vantage point at the Mt. Baldy command center, I was struck by the fact that they were all sacrificing their time to contribute to the search. Some were on their days off, while others had left their jobs to help. Both contributions were equally selfless. Instead of going home and spending time with their families, they were searching for a total stranger in need of help. And the sacrifice extends beyond the team members themselves. Their spouses continually take over extra duties at home while the rescue team member is gone. And their children also are making sacrifices as they wait until morning to see the parent who couldn't be there to tuck them in the night before.

Thank you to all the selfless, courageous people who *never* gave up and to their families for their sacrifice and support of each search-and-rescue team member. Everyone persisted until they found Eric and Gabrielle. We know it takes an army of individuals to conduct a massive search. There really isn't any way to adequately convey our gratitude for their efforts. I can't just say, "Thank you" and expect that to be enough for helping to ensure two people are alive today, especially when one of them is my son. What I can promise is that we will do our very best to pay forward the kindness and perseverance shown by all the rescue teams. And I pray that they will stay safe during all future rescues.

A Few "Whys" of Being Part of a Search and Rescue Team

It is an enormous commitment to volunteer to be part of a rescue team. Each team member has their own meaningful story of why they do what they do. Although we're unable to share the personal stories of everyone who helped find Eric and Gabrielle, we wanted to include a few of them. Thank you to Taison Tran, Alan Leung, and Donovan Leung for sharing their stories. They are the three who planned to stay in the canyon overnight and found Eric and Gabrielle's footprints, which led to the rescue hours later.

(L to R) Alan Leung, Eric, Taison Tran, Donovan Leung.

Alan Leung

I joined Sierra Madre Search and Rescue (SMSR) in March 2017 after I spent a few months researching various teams in the area. This is the first team I've joined. I was really drawn to SMSR because they are an all-volunteer team and cover the area where I grew up and still hike frequently.

When I started, the company I work for was led by a former team member, so they were pretty understanding of the commitment, which was very helpful during the training period.

I really got back into hiking and camping when I moved back home from college. My folks live in Arcadia, just south of Chantry, so I'd go up there a lot. One day I saw the team wheeling someone out and figured since I was up here so much, I might as well help. There really wasn't an aha! moment or a lifelong desire to join search and rescue. In fact, I underestimated just how much time and energy it would actually take, but I also underestimated how much of an impact the whole process would have on me. The bonds you build with your teammates, the satisfaction of helping someone in a very tangible way, or getting a hug from a family member—it is one of the most fulfilling things I have ever done and what keeps me going.

I've had several life transitions since I started with SMSR. I started with a girlfriend and now I have a wife and son. The dynamic has definitely changed. There's more planning involved, and there are some calls I can't respond to because of family commitments. My wife has been incredibly supportive throughout the entire process, but it's not just the two of us now. Luckily, the team is super understanding and everyone shows up in force, so I never feel bad for too long.

My favorite part about search and rescue is the people. Everyone brings something different to the table, but there's a common thread amongst everyone—their love for the outdoors and their desire to help. I can't think of any real downside when working with folks like that.

Donovan Leung

For me personally, the decision was second nature. I've spent a good portion of my life in service to my country and intended to continue that tradition as I moved forward through life. I started at 18 as a U.S. Marine where I served for 12 years and worked disasters such as the Fukushima Nuclear incident, Destructive Typhoon Damian in Philippines, the Thailand coup, and many more. When I transferred to the civilian sector, I searched for a way to still serve, began working for a global crisis response team, and had the honor of doing body recovery in Guatemala for those lost to Volcano de Fuego. However, I wanted to give back to my local community too. Shortly after I began my search, the search-and-rescue team sent me an old-fashioned mail recruitment letter and I applied immediately. Although I was working full-time as a first responder already, it was important for me that I do this, and I haven't looked back since. My wife, who is an ER nurse at LA County Hospital, supports my lifestyle because we share the same passion and interest in serving those in need.

Taison Tran

I have always enjoyed the outdoors and helping people. Throughout college and after, I've been part of various professional organizations, so volunteering is always a big part of who I am and what I do. When Chantry Flats reopened after the Station Fire in 2009, I had a friend visiting and we jumped at the chance to go hiking there. At the trailhead was a sign saying that SMSR was recruiting, and I took a picture of the sign. I filled out an application and have never regretted that moment.

There's plenty of space within SAR to flourish wherever your talents and skills are needed. I was lucky to have a flexible job, so it was easy for me to do some of the bigger searches for out-of-county places like the Sierras or extended searches in Riverside or San Bernardino County. And people go missing in some pretty amazing locations, so there will always be the reward of trying to trace their footsteps and put yourself in the shoes of the person you're looking for.

At the time I joined the team and passed the MRA, I was single and didn't have too many responsibilities, so it was easy to be eager for the pager to go off. Since that time, I've gotten married, completed my MPH, and had two kids. My wife is a PA in a hospital, so she also has irregular hours sometimes, and we both try to be hands-on parents as much as life allows. A lot of my non-SAR time is spent trying to soak up every minute with these kids as they grow up. As a new parent, I need to do a lot of juggling of responsibilities, and my family is my highest priority. At the time of the search for Eric and Gabrielle, my youngest was almost a year old and my wife had the week off from work, so it was a combination of being available and making sure that my kids and wife were okay with me taking a day or two to complete the assignment. I usually read to my kids every night, so it's particularly hard on them when I am not there to help put them to sleep. The same for my wife ...getting two kids to eat sometimes is a multi-person job, and I can appreciate when one person has to tackle two kids. I enjoy SAR so much that sometimes my wife feels like I'm having fun while I'm on a search, so I will always be making it up to her when I get back.

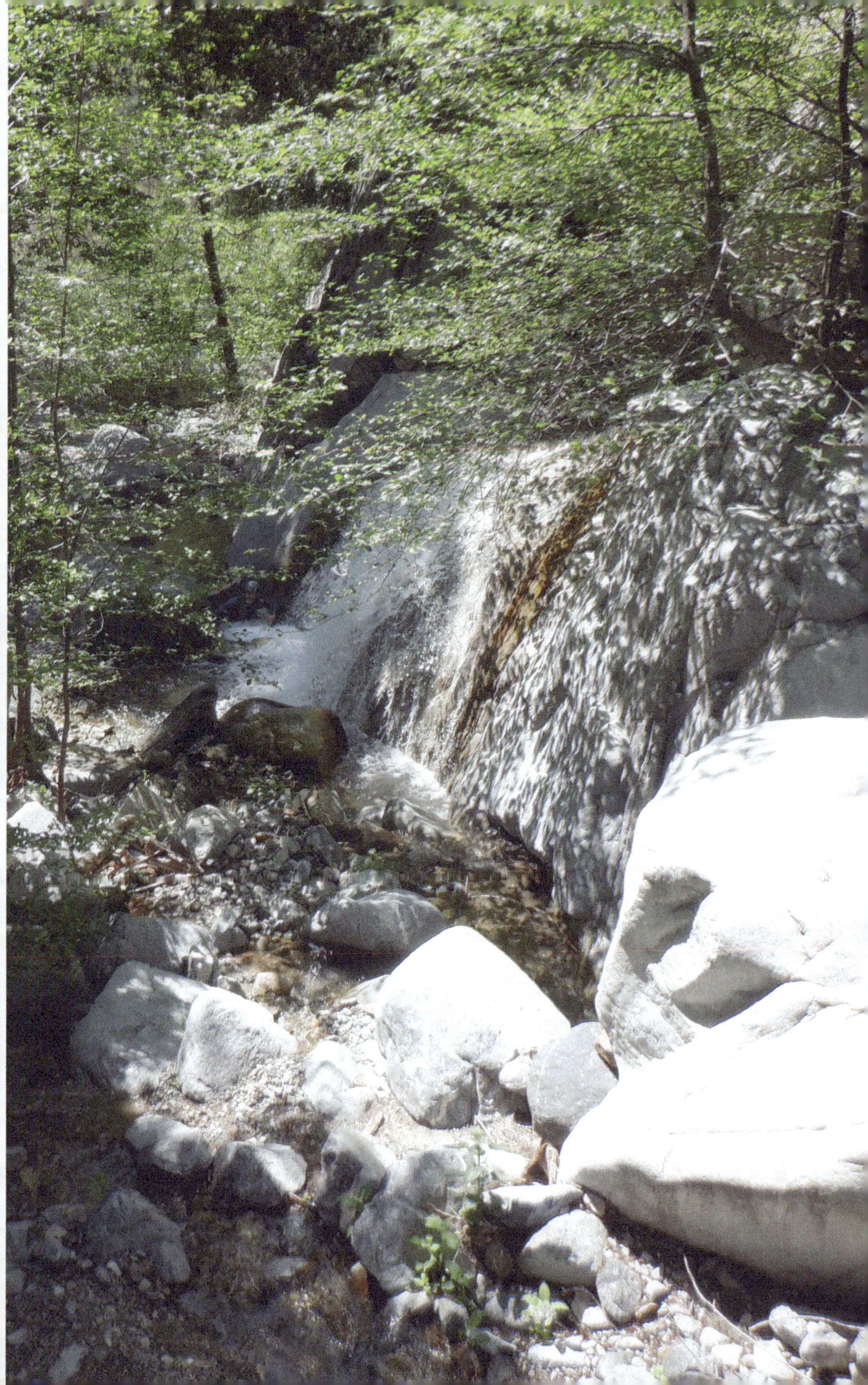

Rescue Crews

Aviation Crew

The first helicopter that Eric and Gabrielle saw that had located them was piloted by Lt. Dan Futscher. He was accompanied by Dep. Cody Korkotsakis. They were alerted that footprints had been spotted in an area, and they flew to that location while there was still daylight and spotted a campfire and two individuals. Confirmation was made through the PA system that the two were indeed Eric and Gabrielle. Due to requiring three crew members to conduct a hoist rescue, the County's second patrol helicopter, piloted by Sgt. Jon Anderson and accompanied by Dep. Greg Ditfurth, met up with Futscher and Korkotsakis at an offsite location. After combining crews, the new crew consisted of Sgt. Anderson as the pilot, Lt. Futscher operating the helicopter's hoist system, and Dep. Korkotsakis performing the role of the rescuer. Dep. Korkotsakis was lowered to the bottom of the waterfall that Eric and Gabrielle were on top of. Unfortunately, the height of the trees where Eric and Gabrielle were located prevented the crew from lowering Dep. Korkotsakis directly to their location.

Due to a combination of challenges, including the approach of nightfall, the height of the trees, and the need to have three crew members onboard to hoist Eric and Gabrielle up, they returned to the Sheriff's hangar located at San Bernardino International Airport.

The crew transitioned into a rescue helicopter, again piloted by Sgt. Jon Anderson who was accompanied by Crew Chief Lt. Dan Futscher, who operated the hoist, and Dep. Greg Ditfurth. The Department's rescue helicopter is a larger Bell UH-1H equipped with a hoist that has 100 additional feet of hoist cable beyond what the Department's patrol helicopters have. The canyon was narrow and the trees an obstacle, and Sgt. Anderson relied heavily on the expertise of his Crew Chiefs to keep him apprised of potential hazards. Dep. Korkotsakis had stayed on the ground with Eric and Gabrielle to prepare them for the extraction, getting them geared with special suits to provide a secure method of transport into the helicopter. Once hoisted up and onboard, they were flown to Mt. Baldy station for medical evaluation.

As with all rescue crews, confidence in the skill level of your team is crucial, as you're relying on each other to keep everyone safe. Pilots and crew members have trained heavily and worked in several areas of law enforcement for years, including patrol duty and incarceration facilities, prior to being part of specialized units. The passion these individuals have for their work contributes to their longevity in their chosen career.

We are grateful to this crew that so skillfully navigated the helicopter into the canyon and got Eric and Gabrielle out safely, and also to all the crews that searched for them throughout the five days they were missing. The determination and skill with which all of the crews worked, on the ground and in the air, was nothing short of phenomenal.

Sergeant Jon Anderson

Lieutenant Dan Futscher

Deputy Greg Ditfurth

Deputy Cody Korkotsakis

Search and Rescue

Thanks to every member of the search-and-rescue teams for your commitment. Your selflessness has saved many, and you will always have our respect and gratitude.

West Valley Search and Rescue

Altadena Search and Rescue

San Bernadino Dog Team

Bear Valley Search and Rescue

China Lake Search and Rescue

Montrose-Crescenta Search and Rescue

Orange County Search and Rescue

Rim of the World Search and Rescue

Riverside Mountain Search and Rescue

Sierra Madre Search and Rescue

RESCUE CREWS

San Bernardino Mountain Search and Rescue

San Bernadino Cave and Rescue Team

San Dimas Mountain Rescue

Victor Valley Search and Rescue

RESCUE CREWS

Wrightwood Search and Rescue

San Diego Mountain Search and Rescue (no picture provided)

West Valley Mounted Posse (no picture provided)

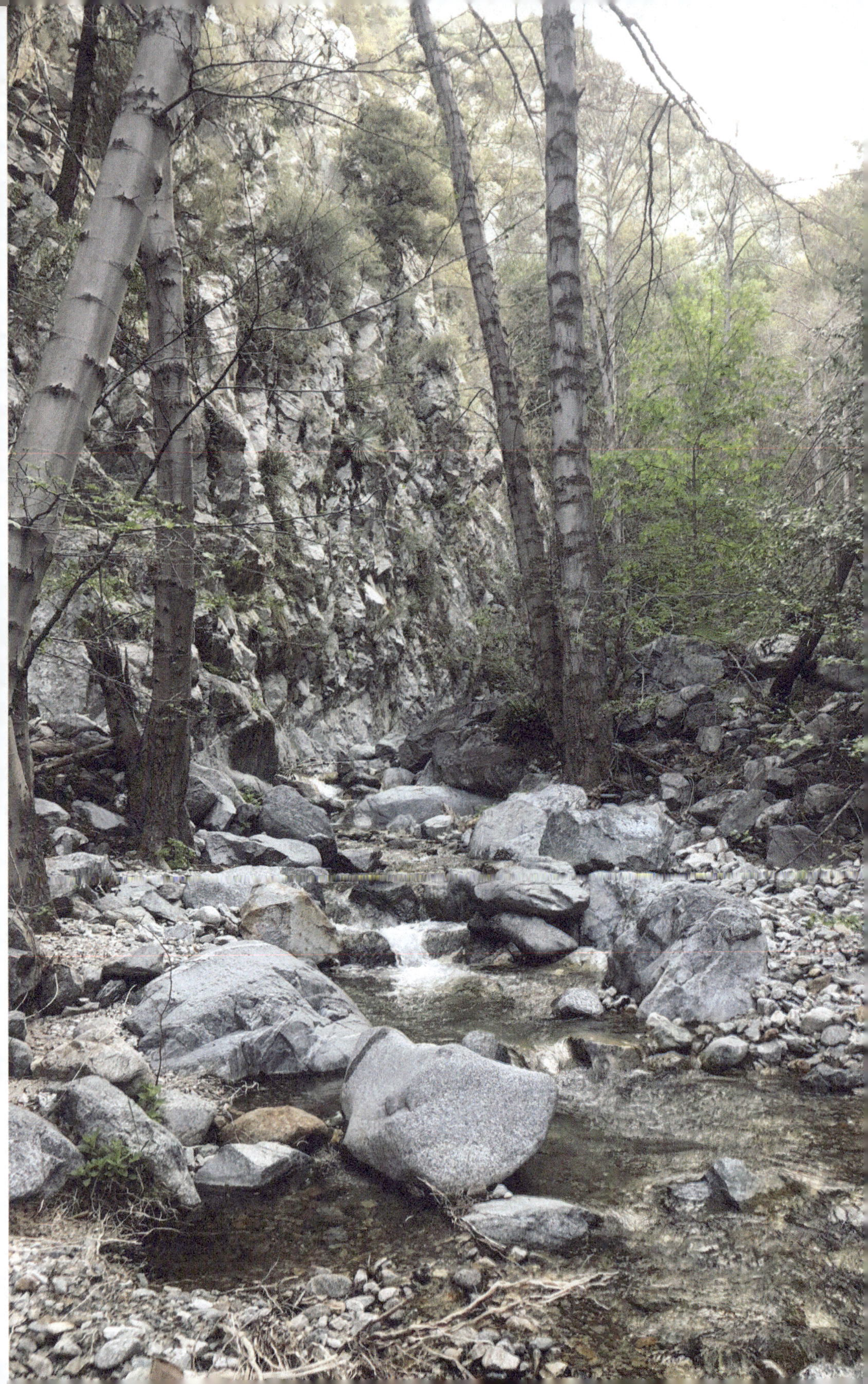

Epilogue

Some people say that everything happens for a reason. I'm not sure about that, but I can say that one thing has definitely led to another in a positive way since Eric's adventure in 2019.

Retrieving Eric's backpack from the canyon took a bit of training. Eric trained with Willie Hunt for a few weeks to gain the skills to safely navigate the terrain with the proper equipment. Willie is a well-known canyoneer who loves exploring mountainous areas off the beaten path. When Eric cold-called Willie to ask for his assistance in retrieving the pack he had left in the canyon, Willie jumped at the opportunity and immediately started map recon to develop a plan while still on the phone with Eric. Once he had taught Eric the required skills to safely make the return trip through Cucamonga Canyon, Willie enlisted his friend Sonny Lawrence as a third member of the team. Sonny is another well-known canyoneer and a member of the San Bernardino County Cave and Technical Rescue Team. Willie and Sonny both made the overnight journey with Eric to recover his pack from the canyon. As a result of his training and the trip through the canyon, Eric discovered his love for canyoneering. This led him to join the San Bernardino County Cave and Technical Rescue Team. Descending canyons is now his favorite pastime.

Rappelling has always looked like fun to me, so I made a trip to California to learn what canyoneering was all about. Although heights have been a mental hurdle for me, I hoped this new experience would help me overcome some of that fear. Eric took me to Little Santa Anita Canyon and Rubio Canyon to learn the ropes. Each time I got on rope, I told myself not to hesitate. That self-talk must have worked, because rappelling was as much fun as I had hoped it would be.

The last day of the trip was spent learning to rappel down a wet canyon with a bit of water flow. We hiked near Lytle Creek to the drop-in point, traversed down to our first rappel, and changed into our wetsuits. The first was a short rappel, about 35 feet, and it was a good place to get accustomed to feeling the power of the waterfall. The second rappel was a bit farther, about 80 feet. So far, so good. I was actually having fun.

When we got to the third rappel, Eric showed me where Willie and he had set a new anchor. I got set to rappel down the 85-foot descent, which would end in a bowl at the bottom. Eric told me to do a baseball slide to get to the edge. This technique keeps the forces on the anchor low and allows for a smooth start to the rappel. The water flow was heavier than the last waterfall, so I felt safer continuing to slide on my side as opposed to putting my feet on the wall. I continued to descend until I was fully immersed in the waterfall. As I was looking at the pool of water below, suddenly the force of the water on my helmet started to push me forward and my head down. With my head being forced downward, the water was now hitting my upper back and forcing me further into a flip. I tried to bring myself upright, but I couldn't. I continued to turn until I was upside down, and the water now was going into my nose and mouth, leaving me unable to breathe at times. I don't know what being waterboarded feels like, but it has to be much like what I was going through.

Moments before the flip.

I tried to grab my whistle to alert Eric that I was in trouble, but my left arm was wedged against the rock, and I couldn't move it. My right arm was keeping the rope taut so I wouldn't fall. To be honest, I think I probably had the rope in the Sqwurel (rappel device) wrapped so tightly that I wouldn't have gone anywhere, but I didn't know that for sure. I got my head up for a brief second to scream for Eric's help. The waterfall continued to pummel water into my nose and mouth, making it impossible to breathe. I thought about taking my chances by letting go of the rope and dropping to the pool below, but I wasn't certain how deep the water was so I decided to view it as a last resort. If I passed out, releasing the rope would have been automatic.

It's not easy to hear anything above the roar of a waterfall. I wasn't sure if Eric had heard my screams or if he had seen me flip. If he had, I knew everything would be fine. If not, I was in real trouble, and I was getting weak quickly. I fought to get upright. I thought "I'm not done yet because there's still so much I haven't finished," and I felt badly that I had wasted meaningful time. As I turned my head to the side, I could see the beauty of the mountains and decided this wasn't such a bad way to go out, having this be the last place I would see.

I was able to pick my head up one more time and look at the canyon. It seemed like I was in such an enormous and magnificent place, and it reminded me how insignificant I was. In a world with billions of people, I would be simply another obituary in the newspaper. But to the people I love, it would be different. I thought about my kids and my grandkids and was reminded that it's been an awesome ride, and I hoped I had left a word or two of wisdom with them for their future. I thought about my friends and how grateful I was to have shared many years with them.

My body was tired from my struggle to get upright and breathe, and by now I was basically a wet noodle hanging by my waist upside down. Then I thought I felt a tug. I wasn't quite positive, but then I felt another one. This was it. Eric had to have heard me or have seen me flip, and he was lowering me down the waterfall. I was happy my arm was free from being wedged against the rock, but it was limp like the rest of me for the ride down. When I got to the bottom, I was sucked under by the swirling water. I struggled to get up, but I knew I didn't want to be tangled in ropes while trying to keep my head above water. I finally got to a place where I could stand and then unhooked myself from being on rappel. The place with the least amount of turbulence was at the edge of the pool, so that's where I headed and where I stayed.

Eric was trying to communicate with me from the top, but I couldn't hear anything. I finally figured out that he wanted me to go back and get the bag. I wasn't confident that I had heard him correctly, but I went back under the waterfall and grabbed the bag and signaled to him that I had it. It took him quite a while to get down to me, but when he did his descent, I thought he looked like Bear Grylls in the TV series Man vs. Wild as he went scaling down the fall, strong and confident.

Safely at the edge of the pool.

When he finally got to me, he said, "What happened?" I explained that I had been drowning, to which he responded that he already knew that. Oh, yeah, yeah …he probably wanted details of why I had flipped, so I just told him I put my head down because I didn't like the sound of the water pummeling on my helmet. Once I looked down, it was all over. The water was then at my back and had a wide surface to continue to force me to flip.

On the next rappel, it was tricky to get the safety tether attached. Well, it was tricky for me anyway. Eric got over to the anchor and clipped in his safety tether. Then he told me to come to him. I told him uh-uh, the water was flowing too fast for my comfort level, and I wanted to avoid being swept over the next waterfall. He said he was tied in and had his leg positioned to stop me from going over the waterfall, so I attempted to swim to him and got pulled under. I tried a second time, got pulled under again, and retreated to the calm area at the side of the pool. I told him to give me just a minute to catch my breath and I'd make it on the next run. He had a different idea. So he threw me the bag and told me to hang onto the rope, and he would grab me when I got close enough. That worked beautifully and he immediately secured my safety line.

Now we just had to get down the next two waterfalls. Eric wanted me to do a baseball slide again to the lip and look over the edge to get a feel for how strong the water flow was. If it made me too nervous to go down on my own, I was to signal him, and he would lower me down himself. I liked the fact that I had options. I got to the edge, took one look over, and immediately gave him the signal to lower me down. The sound of the water on my helmet again throughout the descent was unnerving. When we both reached the bottom, we looked up to see a beautiful waterfall with the sunlight hitting the water.

When we got to the next waterfall, the webbing was a bit sun damaged, so Eric wanted to change it out to be on the safe side. He let me know this would be a much easier descent than the last two and he was confident I could make it down on my own; however, once again I signaled for him to lower me. I guess my confidence had been shattered after the traumatizing experience on the third rappel.

On our way home, we talked about what had happened. We spent the rest of the day running a couple of errands and getting dinner, and then we spent some time chatting. I would fly home the next day. But when I finally tried to get some sleep, I was wide awake until the sun came up and it was time to get ready to leave. I was humbled by what had happened and what could have happened. Had I flipped just over the lip where Eric couldn't see or hear me scream for help, he wouldn't have known to react as quickly as he did. I was grateful he had been there with me that day. Nikki, Tim, and I have always agreed that if something goes wrong, Eric is the guy we want with us. This proved true for me that day. He literally saved me from drowning and got us through the rest of the run safely so I could see another day.

So maybe everything does happen for a reason. What I do know is that thanks to immense prayer, our story of Eric's five days in the canyon ended perfectly. For me, such undeserved kindness from God changed our lives, and we were all able to take something meaningful from it to share with others.

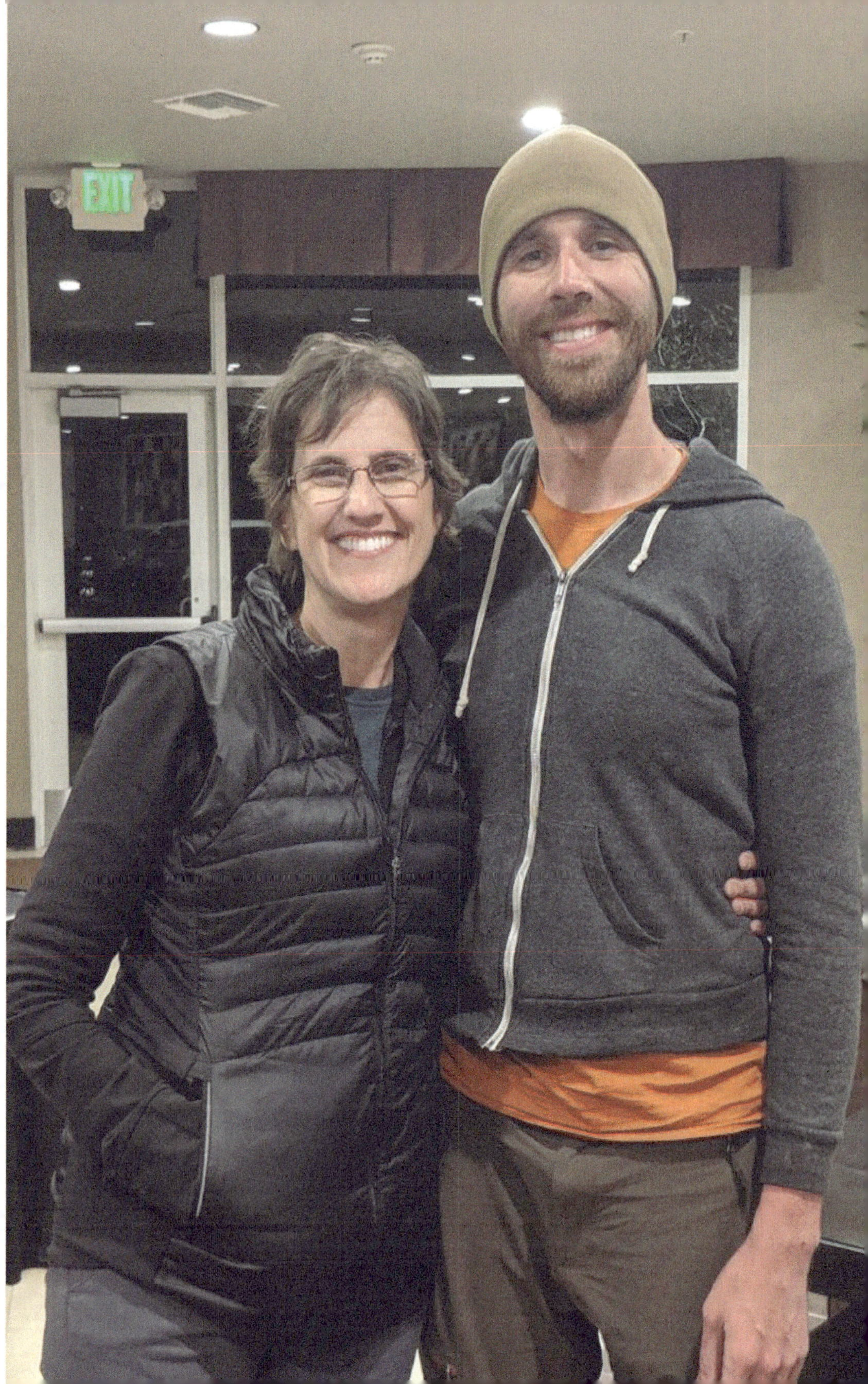
EXIT

Rescue Day

This moment in time will be cemented forever. I'm so blessed to have the gift of this wonderful son in my life. The last few days, I've had memories of him shared with me by others and those that have listened to many of my favorite magical moments. And today I have the gift of knowing that we will be able to continue creating more. Today I have the gift of saying "I love you, Eric" and know that he hears me. There is nothing more that I could ask for.

There were many counties involved in the rescue of Eric and Gabrielle, and the rescue workers, many of them volunteers, are what got us through each day. As soon as we saw them and their prep work began, it gave us hope. These men and women arrived at dawn every day, strapped 60-pound packs on their back, belts with ropes and axes around their waist, and took off into the mountains. As the last crew came in at the end of the day and darkness began to set in, we felt empty and lost. Soon the sun would start to rise, and the rescue workers would be back to give us hope again. There were some rescue workers that worked throughout the night, and that gave us comfort in knowing that Eric and Gabrielle weren't alone out there, and they were the crew that found the footprints that led to Eric and Gabrielle.

Many of these committed and courageous men and women could have been elsewhere that day, and the choice they made was to search for Eric and Gabrielle day after day. Words are inadequate to express my gratitude. They asked for and expected nothing in return, They are selfless, courageous, honorable people, and are the true definition of a hero. I wish I could shake the hand of every single person that contributed to the search and rescue effort. Since I'll never know the names of all of them, the best I can do is put the baton in my hand and assure them that their kindness will be paid forward.

About the Author

Karen Ziebarth grew up in Pipestone, Minnesota, and after several moves to various states over the years, moved to Central Iowa in 1991. She resides in Adel, Iowa, where she runs a small business, enjoys time with her children and grandchildren, and tries to escape to enjoy the outdoors as much as possible. Karen visits national parks yearly with her sons and has a goal of visiting all 63.

If you would like to support the work of rescue teams in the San Bernadino area, email Inland Empire Search and Rescue Council at iesarc-paypal@sbsar.org, or contact any of the rescue teams directly.

ABOUT THE AUTHOR

Karen Ziebarth grew up in Pipestone, Minnesota, and after several moves to various states over the years, moved to Central Iowa in 1991. She resides in Adel, Iowa, where she runs a small business, enjoys time with her children and grandchildren, and tries to escape to enjoy the outdoors as much as possible. Karen visits national parks yearly with her sons and has a goal of visiting all 63.

If you would like to support the work of rescue teams in the San Bernadino area, email Inland Empire Search and Rescue Council at iesarc-paypal@sbsar.org, or contact any of the rescue teams directly.